ROUTLEDGE LIBRARY EDTIONS:
GLOBAL TRANSPORT PLANNING

I0124387

Volume 20

TRANSPORT POLICY IN THE EEC

TRANSPORT POLICY IN THE EEC

JOHN WHITELEGG

Routledge
Taylor & Francis Group
LONDON AND NEW YORK

First published in 1988 by Routledge

This edition first published in 2021
by Routledge
2 Park Square, Milton Park, Abingdon, Oxon OX14 4RN

and by Routledge
605 Third Avenue, New York, NY 10017

Routledge is an imprint of the Taylor & Francis Group, an informa business

British Library Cataloguing in Publication Data
A catalogue record for this book is available from the British Library

ISBN: 978-0-367-69870-6 (Set)
ISBN: 978-1-00-316032-8 (Set) (ebk)
ISBN: 978-0-367-74255-3 (Volume 20) (hbk)
ISBN: 978-0-367-74279-9 (Volume 20) (pbk)
ISBN: 978-1-00-315680-2 (Volume 20) (ebk)

Publisher's Note
The publisher has gone to great lengths to ensure the quality of this reprint but
points out that some imperfections in the original copies may be apparent.

Disclaimer
The publisher has made every effort to trace copyright holders and would welcome
correspondence from those they have been unable to trace.

Transport Policy in the EEC

JOHN WHITELEGG

R

ROUTLEDGE
LONDON AND NEW YORK

First published in 1988 by
Routledge
a division of Routledge, Chapman and Hall
11 New Fetter Lane, London EC4P 4EE

Published in the USA by
Routledge
a division of Routledge, Chapman and Hall, Inc.
29 West 35th Street, New York NY 10001

Printed and bound in Great Britain by
Biddles Ltd, Guildford and King's Lynn

British Library Cataloguing in Publication Data

Whitelegg, J. (John)
    Transport policy in the EEC. — (The
    Routledge EEC series).
    1. European Community countries
    Transport. Policies of European Econom
    Community
    I. Title
    380.5'094

    ISBN 0-415-01258-9

Library of Congress Cataloging in Publication Data

Whitelegg, J. (John)
    Transport policy in the EEC / John Whitelegg.
        p.   cm. — (The Routledge EEC series)
    Bibliography: p.
    Includes index.
    ISBN 0-415-01258-9.
    1. Transportation and state — European Economic Community
countries.   I. Title. II. Series.
HE242.8.W47   1988 380.5'068 — dc19   88-14863

Contents

Figures and tables

Figures

Tables

Preface

The title of this book is *Transport Policy in the EEC*, not 'of the EEC'. This emphasis deliberately conveys the central notion that EEC transport policy will produce results only through the medium of national transport policies, which vary enormously from country to country within the Community. This influence will work in both directions, and it is to be expected that national transport policies will influence the final form and content of an EEC transport policy. Transport policy, then, has to be viewed as an arena for conflict between member states as each strives to ensure that policies and directives maximize self-interest. This creates an unstable environment for policy formulation and one which is fundamentally unsuited to resolving dilemmas in transport. Nevertheless, it is the reality of transport policy in the EEC – and the background against which this book is set.

The book has benefited enormously from the advice and encouragement of others. Thanks are due to Mayer Hillman, John Roberts, Kate Oliver, Sheila Faith (MEP), Susan Hoyle and Don Mathew. All of these gave thoughtful advice and encouragement. In the Department of Geography at Lancaster I have been very fortunate to have the help of a remarkably good team of workers. Pauline Cross, Elsa Drinkall and Lenore Saville all struggled with the text. Peter Mingins and Claire Jarvis contributed their cartographic and design skills, and Sandra Irish her proof-reading skills. All six made Lancaster a very pleasant place to work.

The book was begun in Darmstadt in West Germany and completed in Dortmund in the same country some 14 months later. In Darmstadt I was lucky to have the advice of Peter Sturm and Klaus Schlabbach as well as the facilities provided by Professor Retzko. In Dortmund I have been equally fortunate to work with Helmut Holzapfel. In both places I have learned a great deal about my subject and owe an intellectual debt to the progress which has been made in Germany and made so freely available to me.

J. Whitelegg
Dortmund–Oespel February 1988

Glossary and abbreviations

Cabotage

In its original maritime context it means trade between two ports in the same country. It is now used to describe the ability of road haulier from country A to operate within country B if this is a destination country or is en route to a third country. This implies the opening of road haulage business in any European country to the operators of any other member state.

Harmonization

The adoption of legislation by Community institutions that is designed to bring about changes in the internal systems of member states. These changes are intended to ensure roughly approximate conditions in all states and to contribute to an efficient common market.

Liberalization

The removal of barriers to trade which may operate in one or more member states, e.g. removing any restrictions on the quantity of pricing of transport services which might discriminate against non-national operators, freeing road haulage from controls on quantity and permitting access to markets and transport services throughout the community without distortions which might be imposed by national governments.

Abbreviations

CTP	Common Transport Policy of the EEC
ECMT	European Conference of Ministers of Transport
ECU	European Currency Unit
EIB	European Investment Bank
ERDF	European Regional Development Fund
IMP	Integrated Mediterranean Programme
ONSER	Organisme National de Sécurité Routière
PSO	Public Service Obligation
NCI	New Community Instrument
SIT	Specific Transport Instrument
TEU	Twenty Foot Equivalent Unit (referring to container sizes)
TIEP	Transport Infrastructure Experimental Programme
UIC	Union Internationale des Chemins de Fer
VDA	Verband der Automobilindustrie E.V.

Railway operators

BR	British Rail
CH	Greek Railways
CP	Caminhos de Ferro Portuguese (Portuguese Railways)
DB	Deutsches Bundesbahn
DSB	Danske Statsbaner (Danish State Railways)
FS	Ferrovie dello Stato (Italian railways)
JNR	Japanese National Railways
NS	Nederlandse Spoorwegen (Dutch railways)
NSB	Norwegian State Railways
SBB	Schweizerische Bundesbahn (Swiss Railways)
SJ	Sveriges Statens Järnvägar (Swedish railways)

SNCB   Société Nationale des Chemins de Fer Belges (Belgian Railways)
SNCF   Société Nationale des Chemins de Fer Français (French Railways)
VR   Finnish State Railways

1 Transport policy in the EEC

A book on transport policy in the EEC must overcome many hurdles if it is to present a lucid account of such a large area of policy, and over so extensive a geographical space. The purpose of this chapter is to establish our guidelines and to place transport and transport policy into an EEC and European context.

Transport is such an important part of everyday life, and so much a part of other major policy areas, that it is difficult indeed to draw neat boundaries around it or subject it to orderly dissection. At a national level this presents policy making with serious difficulties, but at an international level the problems are magnified. Within the EEC the Common Agricultural Policy (CAP) has given rise to much discussion, debate, and dissent, but nevertheless it has produced on the ground tangible effects, having an enormous impact on the structure of agriculture and its economic importance in member states. The same cannot be said for transport which is also a common policy within the meaning of the legislation establishing the EEC. Much of this book will be concerned with unravelling the different elements of the Common Transport Policy (CTP) and interpreting its emphases and gaps.

Transport policy in most European countries, and particularly member states of the EEC, is an amalgam of many years' historical evolution tempered by some distinctively different national approaches to issues like regulation and state intervention, railways, and social concerns. Enormous problems are involved in producing a lucid analysis of transport policy in any one nation state which would make an attempt at synthesizing transport policy over the twelve states significantly unproductive, a factor which is not unrelated to the failure of the CTP to take off in any way that bears comparison with the CAP.

Transport policy is most recognizable when states take action in areas of finance, infrastructure, or competition regulation for particular modes, yet these are not necessarily the areas of policy which will most

1

influence the form and structure of transport or contribute to the quality of life and ease of movement of citizens of the EEC. In practically all European countries there is a geographical division of responsibility in transport matters with the state providing the general framework, but leaving a considerable latitude for local interpretation. In the UK the shire counties and recently abolished metropolitan counties represent the local level within which there would be considerable latitude for transport policy innovation, particularly in land use and fare levels for local transport. In West Germany the federal system gives added weight and significance to regional variations in policy, while a third tier at the level of the city or the *Kreis* adds to the diversity. Under these circumstances, transport policy cannot be understood without some knowledge of the interplay between local and national initiatives, the geographical allocation of power, and the diversity of policy initiatives within any one state whether federal or not.

Of increasing importance to an assessment of transport policy within an administrative system is an understanding of land use and environmental considerations. In transport planning it is now commonplace to be reminded of the interdependence of transport systems and land use systems - but this acknowledgement is rarely translated into physical results for the benefit of residents or workers faced with unacceptable traffic environments or long, tiring, and expensive commuter trips. In most European countries traffic planning still takes place against a background of forecasts of greater car ownership and the assumed advantage of new roads, while public transport struggles with a land use system which has not been designed to maximize the advantages of collective forms of transport. Although the results of this kind of planning do vary through Europe, the effects are not dissimilar - road traffic accidents, polluted environments, loss of agricultural land and forest to new motorways, poor travel opportunities for those who do not own a motor car, and environments which are unfriendly both to the very young and elderly. These are transport policy matters. Sometimes they will surface in the shape of proposals for cycle ways or pedestrian areas in cities or in plans for residential areas to reduce traffic speed, or eliminate traffic altogether, but the central question of traffic in society and the role of the private motor car is rarely confronted. Transport policy, therefore, operates through proxy variables, which sometimes produce results but are more often merely cosmetic. It must follow the dictates of wider economic planning, as in the case of the French support of high-speed railways; and in all countries policies which might impinge on ownership and use of cars take a secondary role in importance to the car industry itself and its employment role. In Britain support for the 'company car' (a state-subsidized car for higher socio-economic groups) institutionalizes support for the car and does little to negate whatever progress can be made in public transport with lower fares or the occasional environmental improvement or cycle way project.

Throughout Europe there are many different kinds of approach to land use and environmental improvements which impinge on transport. In Holland, the Woonerfen schemes are a notable example of the kind of drastic improvements which can be produced in residential areas by reducing traffic speed and tipping the balance in favour of children and residents. In Germany, *Verkehrsberuhigung* experiments have had a similar effect and have raised the general consciousness over both what is possible and what is acceptable.

Land use is a key issue in any social or economic planning. Within a national state it will influence the demands made on transport infrastructure and hence on energy consumption and public expenditure. These are real costs which have to be met and which, in their turn, generate further costs in the form of disbenefits to groups of residents and transport users. It is paradoxical, therefore, that land use planning plays such a small part in transport policy. This applies at all geographical scales. Decisions on where to build new schools, hospitals, or workplaces (or which to close) are rarely taken with transport considerations in mind. Regional planning within a nation state has not been effective in reducing the pressure on heavily utilized infrastructure in growth areas while re-deploying such pressure to areas of high unemployment and/or underutilized infrastructure. In fact the opposite seems to be the case, with most European countries still experiencing a regional problem of some kind, together with great pressures in areas like the M4 corridor to the west of London. Growth areas, with their heavier dependence on motorized transport and air traffic than the growth areas of a previous technological revolution, impose particularly heavy burdens on transport and the traffic environment. They are also more likely to be involved in international traffic movements. For the EEC the growth of international traffic (both passenger and freight) has been a major stimulus for increased activity in the sphere of liberalization and harmonization, always with an emphasis on removing barriers and easing movement.

The combined effect of land use changes at a European level, producing areas of high growth and an atmosphere of liberalization at the supranational level, is to stimulate transport with consequential effects on quality of life and the environment for those who experience the increasing levels of traffic. These transport problems are in reality only a special case of a more general problem which is one of coping with the environmental and social consequences of economic growth. Economic growth, when it occurs, is very uneven in its spatial and social impact, generating considerable disbenefits for some groups. An alternative to economic growth is a policy based on conservation and social development. Whether or not the latter would constitute a fair exchange for jobs and other benefits of economic growth is difficult to say, and very much depends on who is gaining and who is losing. What is clear is that transport policies cannot be viewed in isolation from land use changes and economic changes; there is a fundamental relationship of mutual dependence which has to be reflected in policy making at any level, and also one with the creation and alleviation of environment dereliction. Transport policy making at both the national and EEC level has not reflected these basic characteristics of land use and transport systems, but they will nevertheless be emphasized in this book.

This book, then, is about transport policy within the EEC, and the focus derives from the importance of this administrative agency and its attempts to restructure and reorganize major features of social and economic activities. However, it should be remembered that the EEC itself has no overriding geographical logic or spatial cohesion. It was a group of six nations in 1957, and enlarged to nine in 1973 and ten in 1981. With the accession of Spain and Portugal in 1985, it became a group of twelve. Negotiations are currently in progress with the aim of adding Turkey, a further severe test of the EEC's geographical logic. Still this leaves out large and important areas from the net, particularly the Scandinavian countries, and more important from the point of view of transportation, Austria and Switzerland - which must be traversed in Community links with Italy. Here it is not the intention to enter into a debate on the definition of *Europe* merely to note that the EEC is not Europe, and Europe is not the EEC. In practical transport terms this makes life very difficult for EEC-level policy making as major areas of policy, particularly that of international road and rail haulage, will require the co-operation and agreement of non-member states. This is even more the case for inland waterway travel. Freight traffic on the Rhine is of central importance to the international movement of goods and to the rates which apply by road and rail in this highly competitive environment, yet this aspect of European transport is controlled by pre-EEC international regulations and is not easily influenced by EEC policies.

The Rhine is controlled by an international authority with its own treaty obligations, a factor which Despicht (1969) believes to be very important indeed: 'The practical result is that in the middle of the EEC there is an international authority which governs the most important single constituent of West European inland transport but which is not subordinate to the EEC Institutions.'

The geographical extent of the EEC together with the range of different agricultural and industrial economies which it encompasses will also create enormous problems for the EEC-level policy maker motivated by the removal of barriers to trade and having some common framework of regulation to facilitate the movement of goods. The present book cannot describe these varying circumstances in twelve member states or discuss the degree of convergence or otherwise between transport need and policies, but we shall explain the structure of transport in Europe and put into context the relative importance of different modes. This is the basic information which will be needed to evaluate argument about European transport policy and the success or failure of measures designed to advance the Common Transport Policy.

2 The Common Transport Policy

The Common Transport Policy (CTP) of the Commission of the European Communities is one of the three common policies specifically mentioned in article 3 of the Treaty of Rome, the other two being external commerce and agriculture. Transport has a special status as one of the foundations of the Community (like agriculture) and was seen at the outset as both a means of achieving European integration and of accelerating economic development. Later it became equally significant in thinking about regional development, and transport expenditures came to figure prominently in the activities of the European Regional Development Fund and the European Investment Bank.

From the outset, transport policy in the EEC has been based on a series of implicit assumptions which have not gone unchallenged in recent years. These may be summarized as:

1. Transport infrastructure improvements will generate economic development benefits
2. Freeing the operator from restrictive regulations and fostering free competition will improve efficiency
3. Improving journey times is the same thing as improving transport provision.

The first of these assumptions has been seriously challenged (Whitelegg 1985; Vanke 1986) and there seems to be little reliable empirical evidence to support the hypothesis that improved transport will generate new economic development. The second assertion depends upon the ability of operators to translate transport improvements into organizational changes which are then capable of benefiting the consumer and/or reducing the demands made on infrastructure and the environmental costs of, for example, lorry traffic. What is known about operators and their perception of costs (Bayliss and Edwards 1970) does not lend support to this view. The third assumption reflects a fundamental misconception of transport and the related activities which still underpin EEC transport policy; it has remained remarkably resistant to recent assessments of the wider role of transport in the economy and society (TEST 1986; Plowden 1985). The underlying philosophy of the CTP is one which reinforces the importance of free market mechanisms and the narrowest of views about operator-oriented costs and benefits, together with a poorly developed sense of transport's links with the wider economic and social environment. EEC documentation provides ample evidence for this viewpoint:

> So far the CTP has mainly concentrated on the activities of transport operators, both private and public. The objective has been to set free these operators as far as possible from restrictive regulations, to abolish discrimination, to allow free competition and only to create community rules where the proper functioning of the transport market makes them absolutely necessary.
> (*Bulletin of the European Communities* Supplement 8/79, 'A Transport Network for Europe Outline of a Policy', p. 5, para. 3)

More explicitly, while discussing railway policies in Europe, the Commission advanced the following view:

> there is a good case for recommending to member states that the extent of public service obligations of certain transport activities should be reduced or abandoned. In particular there seems to be no reason why goods transport should not, as a general rule, operate on entirely commercial lines.
> ('Progress towards a Common Transport Policy: inland transport', Communication from the Commission to the Council, COM (83)58, 5.2.9, p. 22)

There is little room in this view for an assessment of the effects of heavy goods vehicles on the residents of urban areas, or of Commission proposals for lorry bans and reduced weight of lorries in all member states. The perceived short-term advantages of motorized transport will invariably oust the longer-term advantages of collective forms of transport (e.g. railways) if narrowly based economic arguments are given such prominence. Perhaps more important from a Community point of view, there is little evidence that this approach will lead to more economic benefits than disadvantages.

The CTP has evolved through several metamorphoses in its thirty-year life since 1957, and in spite of its strong ideological base in free market thinking, it has shifted ground in response to pragmatic and political considerations. Despicht, in an early and perceptive study, has put this another way:

> Despite the persistence of certain ideas and practices, community transport policy has been anything but monolithic. It has proceeded by a series of compromises and of changes of tactics and priorities. (1969:21)

Despicht identifies two phases of development up to the time of writing, in 1969: (a) the 'initial phase', in which EEC institutions acted as if they believed that a Community-wide transport system could be introduced by direct assault, and (b) the 'present phase', in which a single Community executive is trying to reconcile the different juridical powers and to overcome the impasse of previous policy making. The dominant theme is that transport policy must keep pace in promoting the structural changes necessary for economic growth with the other major policies of the Community. This policy stance is not entirely clear, but it does show how the CTP is susceptible to major shifts of emphasis, and how in fact its own objectives in the Treaty of Rome are unclear. Despicht identifies the source of some of this uncertainty as the need to clarify Treaty mandates.

The Treaty of Rome established the framework within which the CTP was to be implemented:

(a) article 74 requires the member state to pursue the objectives of the Treaty as regards transport within the framework of a CTP
(b) article 84 restricts the automatic application of this common policy to road, rail, and waterway, but empowers the Council of Ministers to add marine shipping and aviation as they think fit, provided that they act unanimously
(c) article 75 contains the substantive mandate for the common policy; it lays down the institutional procedure to be followed, and stipulates certain measures that must be taken by the end of the transitional period.

These ground rules for the operation of the CTP leave considerable room for manoeuvre in fleshing out the elements of a policy. This is the task of the Commission, who make proposals to the Council of Ministers; and the first such proposals were made in 1963.

The specific injunctions of the Treaty (see table 2.1) are limited in scope and come directly from the experience of the original six members largely in the area of international road haulage. The first major document which clearly set down principles and guidelines for action was the Schaus Memorandum of 1961 (EEC Commission 1961 quoted in Despicht, 1969, p. 33), named after M. Lambert Schaus, the first Transport Commissioner. The Memorandum, which was followed by an Action Programme in 1962, was not a formal proposal with fixed ideas for transport. Its purpose – as explained at the time – was to initiate a wide-ranging

Table 2.1   Summary of specific Treaty articles which guide the CTP

75(1)(a) and (b)   requires common rules for frontier-crossing traffic within the community and for the interpenetration of national transport markets

76   forbids member states to discriminate against foreign carriers (who come from other member states) any more than they did when the Treaty of Rome came into force

79   bans discrimination in transport charges within two years where the condition of carriage and cargo are identical and subsequently where the transport operators are equivalent in practice

81   requires the charges levied at frontier crossings to be related to actual cost (instead of to fictitious terminal charges) and provides for the actual costs of crossing frontiers to be reduced progressively

77   permits state subsidies for the co-ordination of transport or as compensation for public service obligations

78   states the general principle that any measures concerning transport rates and conditions shall take account of the economic circumstances of carriers

80   prohibits specially favourable transport charges (support tariffs) for the benefit of particular undertakings; it contains an exception for regional policy goals

82   permits state aid in transport in the 40-km zone on the West German/East German border

Source: Despicht (1969).

discussion of the options open to a CTP; nevertheless, the Memorandum was visionary:

> A transport policy applying to the transport of the community as a whole must, therefore, gradually replace national transport policies. In the very process of economic integration, the difference between national and international transport within the community will disappear.

Subsequent events were to show that this was both over-ambitious and narrowly based on a particular view of transport and economic development.

Button (1984) has summarized the way in which 'the nature of a transport problem was perceived by Schaus, the principles to be followed and the major policy action which resulted from it'; this is reproduced as table 2.2. The programme of actions formed the Action Programme of 1962 and was to be implemented according to a timetable covering the transitional period of the EEC to 1970.

Table 2.2 The 1962 Action Programme

Liberalization of the rules giving access to the transport market; carriers were to be given new opportunities to supply their services across national frontiers and discrimination between member states and between different modes of transport was to be eliminated; these measures were directed at road haulage. A proposal in 1963 from the Commission put forward a 'Community quota' scheme to permit road hauliers to engage in any frontier-crossing traffic anywhere in the Community. Associated with this was the absorption (i.e. phasing-out) of all bilateral quotas and the eventual extension of the Community quota licence to permit holders to engage in cabotage, or plying for trade within a national territory

Organization of transport prices via the forked-tariff. This is a pricing system where a minimum and a maximum is specified for charges and between these boundaries the carrier is free to fix a charge. This system collapsed in 1966, presumably as the results of relying on price mechanisms were not very good

Harmonization of technical, social, and fiscal regimes. This is an area that is most generally perceived as an EEC concern and, in the case of transport, covers weights and dimensions of road vehicles, conditions of work in road transport, and taxation of vehicles

---

Despicht (1969) is lucid in his treatment of the emphases of the CTP up to 1969. He adopts a functional approach, and identifies three main groups of measures which he claims emphasize the persistence of the subject-matter over time:

1. The application of specific rules on discrimination
2. The promotion of new procedures for co-ordinating investment in transport infrastructure
3. Measures designed to bring about a positive re-moulding of the structure of transport services in the EEC as a whole; these fell originally into three groups - liberalization, organization, and harmonization.

Discrimination

Throughout EEC involvement in transport there is a consistent deference to the principle of 'natural' market forces which are evocative of the Victorian reliance on natural laws as the ultimate arbiter of economic processes. Banning discrimination means the removal by law of anything which 'artificially' distorts the natural condition of a competitive market. Activity in this area pre-dates the EEC and was a major preoccupation of the European Coal and Steel Community (ECSC). The basic principle at stake in both ECSC and EEC strategies (they merged their executives in 1967) is that member states should not discriminate in favour of their own carriers and against those of other states as far as international freight transport was concerned, and that as far as possible transport rates should be published and widely available, so that everything was 'transparent'.

The ECSC took action on railway rates to reduce or eliminate the effects of crossing a border. Discriminating practices which favoured domestic exporters and penalized foreign imports were abolished. Less success was forthcoming in the area of tariff harmonization, even though this did not imply that all member states should apply identical rates over set distances.

The EEC took up the same themes as we have already seen in the Treaty (table 2.1), especially articles 79 and 80. However, discrimination of a more subtle kind continued for many years to be a problem for the EEC, particularly where there was a regional policy dimension as in the border area of West/East Germany, and in attitudes to regional development in the Mezzogiorno region of southern Italy. In the case of the Saar region of Germany, the German government succeeded in maintaining railway tarrifs at 40 per cent less than normal rates on the ground that a canal linking this region with the Rhine was to be constructed, providing 'potential competition' and thus justifying the reduced rate for the railways. In Germany this highly discriminatory tariff was called 'als–ob Tarife' ('as–if tariffs').

The Saar case shows the weakness of EEC legislative procedures in the teeth of fierce national opposition, but also the ways in which pursuit of 'pure' economic logic can frustrate (potentially) the legitimate social and regional policy measures of national governments. If the logic which underpins discriminatory legislation in the Community is taken just a little further, then it would certainly outlaw much of what poses for regional policy and put into jeopardy any governmental intervention which has been focused on a well-defined geographical area with particular social and economic problems. Also there are clear illogicalities in discrimination arguments, in that they draw an artificial boundary around freight transport rates and ignore the whole issue of transport infrastructure investment, which can be just as discriminatory. Much of Britain's motorway network has been justified on regional policy grounds, as has the Italian motorway expansion. It is debatable whether or not the objectives have been achieved, but the Saar rate discrimination is not so very different from the decision to provide high-quality transport links on Merseyside or in the Ruhr or southern Italy.

Co-ordinating investment

The EEC has not been slow to recognize the importance of intervention in transport infrastructure and in later years this becomes more signifi-cant as an area of transport policy. In 1960 and 1961 the EC Commission produced recommendations which set out the principles on which a network of 'trunk routes of community importance' should be defined and indicating which road, rail, and waterway projects should be given the highest priority. This did not come to much, mainly because of the lack of correspondence between routes which had an EEC significance (e.g. those which crossed frontiers) and those which had a national significance (e.g. those linking urban areas or making disadvantaged areas more accessible). The reaction of member states was lukewarm, and by 1966 the idea had been diluted to the extent that all that remained was a vague consultation procedure which left the choice of projects for study to the discretion of member states.

It would be difficult to co-ordinate investment in a situation where each member state had a different approach to transport taxation and methods of allocating and controlling expenditure. In 1963 the Commission instituted an enquiry to determine (a) the total cost of the road, rail, and waterway infrastructure of the EEC, and (b) the costs that should be apportioned, first, between the transport and other functions of the infrastructure, and secondly, between the different categories of user of the transport infrastructure.

According to Despicht (1969), the Commission had two objectives. First, to impose restraint on the discriminatory subsidies concealed in transport infrastructure investment by ensuring that the costs borne by the state were recouped in full by taxation, and secondly, that users paid fully in accordance with the benefits they derived from the infrastructure. This would put all users on an equal footing and improve the competitive environment. In 1978 the Commission was still struggling with basic issues of infrastructure (see chapter 3) and presented proposals to the Council to improve the consultation procedure instituted in 1966 and set up a Transport Infrastructure Committee. The Commission also asked that Community financial support be made available for certain major transport infrastructure projects (EC Commission, *Bulletin*, 1979). (We shall return to the question of infrastructure later in this chapter in discussing the current status of the CTP.)

The 1962 Action Programme which followed the Schaus Memorandum contained three main groups of measures, which are summarized in table 2.2 - details are not discussed in the table as this is left for later chapters on each of the important modes. Much of the Action Programme was abandoned and, in his summary of the decade 1958-69 of the CTP, Despicht (1969) summarized the achievements of the CTP as:

1. Establishing a procedure of consultation on national transport policies
2. Implementing the Treaty of Rome on discrimination practices
3. Introducing rules of competition for the transport sector, together with a reduced community quota for road haulage and a forked-tariff regime confined to road haulage for an experimental period only
4. Introducing measures to reform the structure of inland transport, in particular for rationalizing railway operations and for a new system of allocating infrastructure costs.

The generally disappointing performance of the CTP and its perception as a weak area of EEC policy has not diminished with the passage of time. Button (1984) makes use of the same points as Despicht (1969) in commenting on the performance of the CTP until 1973, when enlargement of the original six to nine injected some new life into a flagging transport policy.

The three new members who joined the EEC in 1973 (Denmark, the UK, and Ireland) brought with them a different perspective on transport and a lack of familiarity and patience with all the legalistic undergrowth which had surrounded tariff fixing and cross-frontier movement within the original six. This had undoubtedly stunted the growth of a CTP which was more concerned with one or two modes and the competitive environment in which they operated than with general transport problems on a European scale. Whether the three new entrants really brought with them a breath of fresh air, or whether the CTP was played out and ready for a new push in the changed circumstances of a nine-member EEC, does not really matter. The post-1973 development of the CTP shows some signs of life but had inherited too much inertia, caution, and narrowness of outlook to bring about fundamental change.

One of the first changes noted by Button was an extension of the scope of community interest to include air and maritime modes; a judgment by the European Court of Justice in 1974 ruled that neither mode was exempt from the obligations of the Community and that general rules applied to them (Button 1984). Other changes were also evident, especially in the way transport increasingly was discussed in terms of its links with regional, fiscal, industrial, environmental, and energy affairs.

The Commission set out an action programme to reflect the new mood and take account of the new members. This 1973 Action Programme did not depart significantly from its predecessor; it still argued the benefit of competition and emphasized the free circulation of transport services and harmonization of conditions of competition, giving a great deal of importance to the free interplay of market forces. It took the CTP firmly into the area of infrastructure policy and away from the emphasis on operational controls, such as quotas and tariffs:

> The most important ingredient in this modified approach was the emphasis on the integration of the national transport system into a community system which requires community action in the planning and financing of the transport network and in the organisation of the

transport market with the ultimate aim of achieving the optimal use of resources employed in the transport sector. (EEC 1983b)

The programme of action accompanying the 1973 communication emphasized the urgent need to come to grips with:

1. The development of an optimal transport network in accordance with an agreed master plan
2. The imputation of the costs of using the transport infrastructure
3. Defining the role of the railways in the future transport system and solving their financial problems
4. Progress in the development of inland transport markets.

Erdmenger (1983:15) points out that one of the main aims of the Commission in 1973 was to 're-emphasise [that] any integration in transport had to take equal account of the interests of transport areas, transport operators and transport workers'. In this sense, the 1973 communication is different from the 1962 Action Programme, but in the substance of what is actually done the difference is not as great as might be expected from the text of the communications themselves. In fact very little progress of any kind was made and, in 1977, the Council noted the Commission's programme and declared that: 'it would endeavour to take account of these priorities in the future proceedings.' This lack of progress was becoming increasingly irksome to the European Parliament and, in 1978, they threatened to take the Council before the Court of Justice.

In March 1981 the Council did convert some Commission proposals into a Resolution which laid down ten new principal points for a transport programme in the period 1981–3. However, the Council was still very reluctant to adopt decisions on what Erdmenger (1983) calls the Commission's 'thirty-six priority items'. This inability to convert proposals into action and to advance the CTP with an obvious intent to make progress eventually led, in 1983, to the European Parliament taking the Council of Ministers to the European Court of Justice.

In May 1985 the Court of Justice ruled that the Council had infringed the Treaty of Rome by 'failing to ensure freedom to provide services in the sphere of international transport and to lay down the conditions under which non-resident carriers may operate transport services in a member state'. The European Court of Justice judgment (13/83) states that the Council failed to provide the freedom to provide transport services within the time period specified by article 75 of the Treaty of Rome.

The 1981 Resolution which set up ten priority topics is a useful guide to the contemporary CTP and its emphases:

1. Improving the situation of railways
2. Continuing measures to harmonize conditions of competition
3. Implementing measures in the field of transport infrastructure
4. Developing combined forms in transport
5. Facilitating frontier crossing
6. Improving the operation of the transport market, particularly international transport
7. Improving the efficiency and safety of transport
8. Bettering social conditions in the transport field
9. Continuing the work in hand on sea and air transport in accordance with the conclusions reached by the Council in its previous meetings
10. Solving the problems likely to arise in connection with intra-Community transit via non-member countries.

These points continue a well-established CTP tradition of making vague statements about taking an interest in some area or another, and are remarkable for the degree to which they completely miss the relevance of transport to the daily lives of millions of Europeans. Efficiency and safety are linked together, as though they are different aspects of the same problem; and nothing is said or done about the condition of public transport or road traffic accidents, or the environmental damage caused by new roads which are justified by increased goods vehicle traffic activity.

In 1983 the Commission issued another communication (EEC 1983b) with a revised work programme and a list of thirty-seven proposals pending before the Council. Erdmenger (1983:24) says of this: 'In its new policy paper the Commission pursues the same ultimate objectives as the Memorandum of 1961 and Communication of 1973, which regards an efficient transport system as a crucial part of the establishment and further development of the internal market, and as being indispensable to economic growth.'

The record of the CTP is poor. In twenty-five years there has been little development of its basic thinking about transport and much repetition and bureaucratic non-activity which passes for a common policy. Its resilience to popular, academic, and critical transport policy developments is remarkable and it exists in isolation from transport priorities, wrapped up in a language of its own and nurtured by a self-perpetuating and somewhat incestuous relationship with wider EEC goals of economic growth regardless of the consequences.

3 Infrastructure

The early years in the evolution of the Common Transport Policy (CTP) were operator-oriented with a clear bias towards facilitating competition by action on tariffs, banning discrimination, and improving frontier crossings. In the late 1960s and early 1970s a shift of emphasis began to emerge and infrastructure appears more frequently as an item of CTP concern. It would be wrong to suggest that this emphasis was entirely new. There are references to infrastructure provision in the early years but these were not elevated to the level of coherent policies. This emphasis on infrastructure is important because it takes the CTP more forcefully into the public domain, and into the area of direct negotiation with national governments on public expenditure – an area where views on national sovereignty and the ranking of expenditure priorities might well produce more conflicts and inaction than the operator approach did. The shift of emphasis is also important because infrastructure is a critical dimension in regional policies, and in energy and land use policies, as well as having considerable social implications. Infrastructure policies could well have the effect of bringing transport policies into the mainstream of European social and economic development.

From a geographical standpoint, infrastructure policies are particularly important. The selection of modes and of projects for investment will involve a restructuring of spatial relationships in Europe. Certain corridors of movement can be identified as particularly important for Community freight and people movement, and the selection and investment will itself focus more actively on that route. This is likely to bring considerable disbenefits to those areas of Europe which lie across main transit routes, especially those areas where routes already converge on narrow corridors, e.g. the Brenner crossing between Germany and Italy and sensitive valley communities on the Munich-Rome axis. Infrastructure investment linked to economic growth and development may produce unacceptable environmental pressures on sensitive areas,

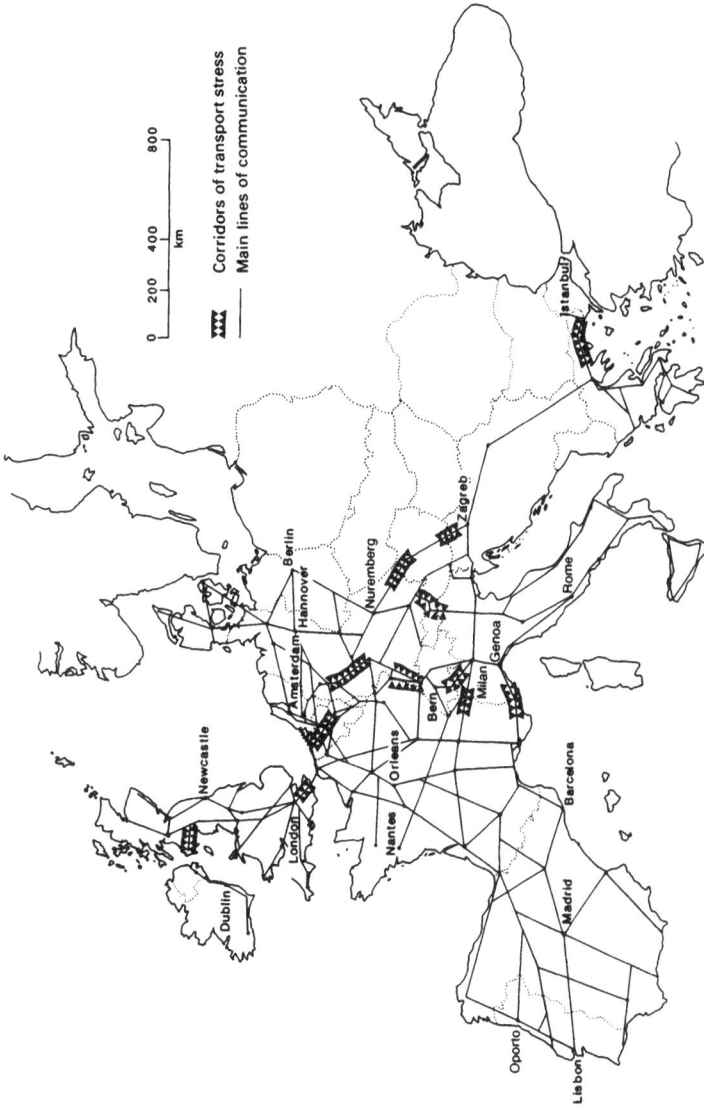

Figure 3.1 Areas of Europe experiencing environmental stress from transport pressures

including those areas of non-EEC countries which experience transit traffic (Austria, Switzerland, and Yugoslavia). Figure 3.1 shows the areas of Europe most likely to suffer increased pressures as a result of CTP infrastructure policy and the growth of road freight.

Disbenefits can arise from the re-allocation of economic activities which result from changed accessibility surfaces and locational adjustments among large manufacturing and distribution firms. These problems were discussed at a conference on the effects of roadbuilding (Whitelegg 1985), and in a recent review of the links between road construction and the urban economy (Vanke 1986). There is considerable evidence pointing to the damaging effects of improved transport systems through loss of jobs as firms rationalize their operations and small- to medium-sized firms in cities lose their competitive advantage to larger firms with lower site costs and now using improved transport links to attack new markets. Improved efficiency in transport and an adherence to competition as a primary goal can, under these circumstances, produce considerable social, economic, and environmental costs.

In a major statement on infrastructure (EEC 1979a) the Commission's view is outlined:

> The Commission has arrived at the conclusion that the CTP will not achieve the objectives defined for it in the Treaty and play its part in the economy as a whole unless it relates more and more to transport infrastructure. The reasons for this new impetus, which originates largely in recent economic development, are clear. (EEC 1979a:5)

The Commission justifies the need for an infrastructure policy by reference to a number of factors. Among these is the increasing international flow of traffic which creates demands of relevance to the EEC but perhaps not so important to one country. This will require EEC-level co-ordination and funding. There is also an assertion that networks are growing more interdependent, thus making national-level policy less relevant than it used to be.

The Commission recognizes that transport investment has economic and social consequences, and states that in the period prior to 1979, 'a disproportionate priority has been given to the development of roads and motorways'. It also recognizes that investment may have negative environmental consequences on the regions crossed by the routes, but offers no guidance on how these important concerns are to be reconciled with the major goals of the CTP and their emphasis on economic growth and integration. As in previous statements of principle or intent, the actions and effects do not bear a close resemblance to the text.

In 1978 the Council adopted the first proposal which became the decision of 20 February 1978 instituting a consultation procedure and setting up a committee in the field of transport infrastructure. The results of this were twofold: member states were to communicate to the Commission their projects and programmes for the development of transport infrastructure and projects of Community interest; these would then be discussed by member states; and a transport infrastructure committee would be set up under the auspices of the Commission to examine every aspect of the communications network of interest to the Community.

Progress with these measures was not particularly encouraging. In 1983 the Commission reviewed progress:

> the Commission itself has completed a number of studies aimed at identifying infrastructure requirements as seen from the point of view of the Community and at developing criteria for the determination of the Community interest in infrastructure projects. But the Council has yet to take a decision on the key questions of complementing its development of national transport infrastructures by a Community dimension, of co-ordinating more effectively national planning in the interests of developing Community networks and of the planning evaluation and financing of specific projects of Community interest. (EC Commission 1983b:11)

This lack of progress does not do justice to the efforts of the Commission to come to grips with infrastructure priorities. In its 'network policy' document (EEC 1979a) some details are presented on the precise, physical objectives of the infrastructure policy. A number of different categories of investment were identified:

1. Bottlenecks likely to hinder traffic between member states
2. International links between major centres (e.g. Brussels–Cologne, Utrecht–Cologne–Frankfurt, Amsterdam–Brussels–Luxembourg–Strasbourg)
3. Links with peripheral regions (e.g. Dublin–Belfast–Londonderry, Dublin–Cork/Galway, and links with East Anglia in the UK and with the Mezzogiorno in Italy)
4. Links affected by the accession of new member states, by land, sea, and air, to take account of the expected increase of traffic following the accession of Greece, Spain, and Portugal

5.   Links overcoming natural obstacles (e.g. the Channel Tunnel, links
      between West Germany and Denmark (via Fehmaru), the Alpine
      links between West Germany and Italy, and the Apennines
      crossings)
6.   'Missing links' between existing networks, particularly for
      motorways between Belgium and France and between the North Sea
      and the Mediterranean via the Rhine-Rhône Canal; for the motor-
      way network the link Thionville-Luxembourg-Trier is identified.

In 1982 the Council asked the Commission to prepare 'a balanced and
experimental programme extending over a 3–5 year period, comprising
precise infrastructure projects': the projects should be able to be
completed, or almost completed, during the period 1984–8 and should be
consistent with previous decisions on bottlenecks, finance, definition of
community interest, together with a strange reference to the results of
consultation on road projects in the Grand Duchy of Luxembourg. All
member states were invited to submit lists of projects for consideration,
and these give a good indication of investment priorities in transport as
they are perceived by the member states. As in other areas of EEC policy,
there is an element of 'gamesmanship' in the lists as governments try to
anticipate Commission thinking on the criteria for support. Nevertheless,
the projects are a guide to modal preferences, areas likely to come under
greater stress, and national perceptions of priorities. The list is reproduced as an Appendix (see page 202). The total costs of the projects
submitted broken down by mode are shown in table 3.1.
     The projects in the Appendix represent a wide selection of major
improvements to transport infrastructure and fall within the general
framework of transport investment based upon reducing user cost and
facilitating the movement of larger volumes of passengers and freight.

Table 3.1   Costs of projects in experimental infrastructure programme (in
million ECUs)

Rail	2,500
Road	5,500
Inland waterway	980
Ports, airport and air control	700
Total	9,680

Table 3.2 Projects supported in first phase of experimental programme, 1983–4

Year	Location	Type of project
1983	Athens–Volos–Evzoni Greek/Yugoslavian border road axis	Improvement of sections of the route between Volos and Evzoni (complementary intervention to that foreseen within the framework of a limited Regulation proposal in the field of transport infrastructure); expenditure of 10 million ECU from the 1982 budget in part
	Rosslare–Dublin towards Belfast (Northern Ireland) road axis	Improvement of this route, in particular the construction of by-passes
	Rotterdam–Cologne–Stuttgart (the Netherlands–RFA) rail axis	Various improvement projects of the capacity of certain sections and of the installation of combined transport and trans-shipment facilities on this axis
	NW–SE transit route (Austria)	Project to be specified through the current negotiations with Austria on transit questions
1984	UK–Continent via East Coast ports	Electrification of Colchester–Harwich rail line; improvement of the port installations at Felixstowe and Harwich
	NS rail axis (Copenhagen–Frankfurt–Milan) (Denmark–Germany–Italy)	Various projects to improve capacity of some sections and installation of trans-shipment facilities
	Luxembourg–Trier road axis (Luxembourg–RFA)	Construction of sections of motorway, in particular missing border links
	NW–SE axis	Projects to be specified (rail, road, ports)
	Inland waterway link between Belgium and the Netherlands	Modernization of the Zuid–Willemsvaart Canal

Source: EC Commission, COM (82) 828 Final, 10 December 1982.

This is in tune with the CTP as it has developed so far, but still fails to reflect wider transport concerns and the rhetoric about environmental problems and quality of life which has crept into EEC documentation in the second half of the life of the CTP.

The cost of the projects was much greater than the resources available (150 million European Currency Units for the first two years of the programme), so it was obvious that only a small number of projects could be supported and these at only a very small percentage of the total cost. The projects supported are shown in table 3.2. In addition, funds were to be made available for 'preparatory technical works for the construction of a fixed Channel link' and for a feasibility study for a Train à Grand Vitesse (TGV) rail link along the Paris–Brussels–Cologne axis.

In the second phase of the expenditure programme (1985-7) the Commission proposed to pay particular attention to projects which will enable rail to realize its full potential and which will improve connections between maritime, air, and land modes of transport. The problems of new member states and 'third countries' (e.g. Austria and Switzerland) would also require some attention.

In 1986 the Commission spelt out in more detail its objectives for a 'medium-term transport infrastructure policy' (COM(86)340 Final). These objectives include:

1. Improvement of transport communications in 'land–sea' corridors
2. Reduction of the costs inherent in transit traffic, in co-operation with non-member states concerned notably by works designed to support combined transport
3. The integration of the peripheral regions into the Community's network
4. The construction of links offering a high level of service between the main cities, particularly high-speed rail links (160 km/h and over).

The 1986 Commission document makes it clear that the modernization of ports and airports is an integral part of the infrastructure programme, and particularly so because of the need for economic development in the peripheral and isolated regions which are more dependent on these modes. The report also goes some way to clear up the confusion surrounding the financing of transport infrastructure projects by suggesting the 'concentration of budget resources' for this purpose. These currently include ERDF, IMP, EIB, NCI, and SIT (see Glossary), and the Commission proposes to use SIT in particular to allow

Table 3.3 Priority investment projects in medium-term programme

(a) Projects situated on axes carrying major community flows and involving heavy burdens for transit countries:
  - access route to Mont Blanc Tunnel
  - strengthening of rail axes
  - improvement of routes giving access to transport links, e.g. over the Brenner or Tauern passes
  - modernization of links in waterway network, e.g. Seine-Escaut-Rhine-Rhône

(b) Improvement of land-sea links, particularly in the corridor UK-Ireland-Continent:
  - rail and road access to Channel Tunnel
  - development of ports North of the Thames
  - improvement of North-South links in Ireland
  - research with new technologies of relevance to Strait of Messina, Gulf of Corinth, Scandinavian links

(c) Progressive improvement of links of a high level of service between the capital and major urban areas, especially high-speed links:
  - Paris-Brussels
  - Cologne-Amsterdam

(d) Better integration of remote regions into the Community network:
  - NW-SE axes serving Greece
  - East-West link to facilitate transit in Greece (Igoumenitsa-Volos)
  - NE-SW routes serving the Iberian peninsula
    Irun-Madrid-Algeciras (road and rail)
    Barcelona-Madrid-Lisbon (road and rail)
    Mediterranean road axis Narbonne-Algeciras
    Road and rail axes Irun-Burgos-Oporto and Irun-Burgos-Coimbra-Lisbon-Faro
  - Improvement of trans-Pyrenian links (Pau-Saragossa) and modernization of railway frontier crossings

Community action to go beyond a simple contribution to pre-existing financial plans. The Commission also envisages the use of private capital 'to the full' to make up shortfalls in Community budgets – as in the case of the Channel Tunnel. In assessing budgetary requirements from within Community resources the Commission is not precise. Projects of Community interest already identified by the member states represent (for ten to fifteen years ahead) an investment of over 20 billion ECUs; the cost of projects already in preparation has to be added to this figure. Table 3.3 shows the Commission's priorities in 1986 for investment, and comparison with table 3.2 shows a greater emphasis on peripheral regions and new member states.

The Commission activities in the field of transport infrastructure and its role in the development of the CTP does not tell the whole story of EEC involvement in transport financing. The EEC works through other policies and with other institutions to the same end. Between 1958 and 1984 the European Investment Bank loaned 3,900 million ECU to the transport sector. Since its creation in 1975, the European Regional Development Fund has given over 3,300 million ECU in grants to improve transport infrastructure. The support of ERDF and EIB for transport infrastructure projects amounted to 5.8 billion ECU between 1981 and 1985 (COM(86)340 Final). Among projects financed by the EIB and ERDF are:

- sections of motorway between Antwerp and Breda, Paris-Brussels, Paris–Metz–Saarbrücken, Lorraine-Burgundy, Bordeaux-Poitiers, Messina-Palermo, the Calabrian and Frioul motorways, the roads from Patras to Olympia, the Breton road improvement plan, and hundreds of sections of the French departmental road network
- airports at Birmingham, Manchester, and Palermo
- upgrading of the Inoi-Larissa section of the Athens–Thessalonika railway line, and the Tyne and Wear metro in the UK
- Irish Sea ferries from Britain to Ireland
- ports including Ramsgate, Harwich, Sete, La Rochelle–Paltice, Boulogne, Calais, and Cork–Ringaskiddy.

The accession of Greece, Spain, and Portugal has created new demands on funds for transport infrastructure investment, but the Community is still opposed to the creation of a permanent fund. Heavy demands are foreseen in the Mediterranean region, 'to remove the Mediterranean from the periphery of Europe and develop its potential as a commercial crossroads between north and south' (EEC 1982:5).

Another area of increasing demands is that of Alpine crossings and links with Italy and Greece via Switzerland. This is the 'third country' concern already referred to, and it is one which is particularly relevant to broader EEC concerns about bottlenecks and cross-frontier movements. Erdmenger (1983) reviews the problems of transit traffic through Austria and Switzerland, including the Austrian unilateral transit tax of 1978, 'which put European transport policy into disarray', and the Swiss limit of 28 tonnes for heavy vehicles using the Alpine passes. It is clearly going to be unacceptable to transit states to take increasing volumes of heavy goods traffic by road arising from the 'success of European integration'.

EEC activity to overcome these difficulties has taken the form of transport agreements with Austria, Switzerland, and Yugoslavia, together with a trade co-operation agreement in 1980 with Yugoslavia. Yugoslavia's geographical position is crucial to links with Greece, and under the co-operation agreement, Yugoslavia will receive aid from the EIB to build a motorway from the Austrian frontier to Belgrade. Also in 1980 the Commission sent a Communication to the Council advocating a financial contribution to the construction of the Pyhrn motorway in Austria.

The CTP's involvement with infrastructure has followed very closely its well-established philosophy of facilitating economic development and the quick and easy movement of freight. The possible range of applications of infrastructure policy has been very narrowly defined indeed to a geographical scale which concentrates on international and inter-regional movement and to an approach which 'must always be to keep the overall economic cost of the transport system as low as possible' (Erdmenger 1983:64). The lack of a passenger infrastructure element at a sub-regional or urban scale, and the lack of a clear view of the relationships between infrastructure provision and the social and economic implications for different groups and different geographical scales, is a major omission and a lost opportunity.

Infrastructure investment and pricing

The relative importance of different modes and different kinds of infrastructure investment is not just a matter of developing a network and treating transport investment in a militaristic 'grand strategy' manner. This is appealing to centralized bureaucracies and can be more quickly converted into physical results than can intervention in the messy area of infrastructure pricing and investment. Not only are there problems of sovereignty with the latter approach, for national governments will not be keen to sacrifice decisions to some 'higher' authority, but there is also a problem of how investment in different modes are compared and how this relates to broader issues about comparability between road and rail investment, prices to users, and decisions on future investment in the light of the perceived economics of different modes.

The EEC is aware of these problems and has tended to take a strong commercial line, particularly with respect to railways:

> one of the objectives of the common transport policy must be to eliminate disparities liable to cause substantial distortions in competition in the transport sector. Measures should accordingly be taken ... as regards state intervention in transport: to reduce PSO's to a minimum; to provide fair compensation for financial burdens resulting from obligations which are maintained and from those involving reductions in rates on social grounds; to normalise the accounts of railway undertakings; to make such undertakings financially autonomous; and to lay down rules governing aids for transport, taking account of the distinctive features of that sector. (EEC May 1965, quoted in Button 1984:45)

The railway operation is discussed more fully in chapter 4.

Despicht (1969) regards the problems of allocating infrastructure costs as an even greater obstacle to the co-ordination of investment than the mention of national priorities. In 1963 the EEC Commission proposed that an enquiry should be undertaken for the Community as a whole to determine both the total costs of the road, rail, and waterway infrastructure of the EEC and the costs that should be apportioned, first, between the transport and other functions of the infrastructure, and secondly, between the different categories of user of the transport infrastructure. Despicht is clear about the reasoning behind these proposals:

> First that restraint should be imposed on the discriminatory subsidies concealed in transport infrastructure investment by ensuring that the transport infrastructure costs borne by the State were recouped in full from transport taxation. Secondly that the initial competitive conditions of carriers would be harmonized by ensuring that the various categories of user paid fully in accordance with the benefits they obtained from the infrastructures. (Despicht 1969:42)

These are important motivations which could, if realized, produce a very different environment for road–rail competition. However, subsequent events do not provide a great deal of evidence to support Despicht's thesis and there has been nothing dramatic in the CTP to bring road taxation in line with the costs of road infrastructure or to disturb the balance of investment preference, so that it more closely reflects the real cost of road provision and the real benefits of rail transport.

In 1965 the Commission announced a series of studies of actual infrastructure costs and benefits in order to be better informed about the issues. A pilot study was conducted of Paris–Le Havre, though Despicht dismisses it as 'confined to empirical data'. In 1983 the Commission declared: 'Although progress has been disappointing since submission of the Commission's first comprehensive proposal in this field in 1971, developments at expert level should allow renewed efforts to be made' (EEC, 1983b:30, para. 5.5.0).

The 1971 proposal followed another study, in 1970, and the establishment of a reporting system for infrastructure expenses and utilization. The proposal suggested the introduction of a system of infrastructure charging for all inland transport modes, based on charging the social marginal cost of using the infrastructure, including accident and congestion costs, as well as the usual track costs. This was a 'railway-friendly' proposal with considerable implication for beleaguered European railways if implemented fully and enthusiastically. This was not to be the case. For vans, lorries, and buses taxation was to be related to the amount of damage or wear-and-tear done by linking it to axle weight and vehicle configuration. This has not come about, and there has been considerable resistance in member states to EEC-wide agreement on vehicle taxation. It is also doubtful if the Commission itself was ever in favour of such drastic changes in the competitive environment of transport. The Commission's general views on freeing and liberating market forces form a natural partner with support for road-based economic development and integration, an interpretation which can be supported by close examination of railway policies (see chapter 4) and by reference to Erdmenger (1983), a former director-general for transport of the EEC, who is ambivalent on this point:

> The European transport policy recognises that the user of transport infrastructure should pay for the resulting costs, even though this principle still has to be put into effect over a large part of the field of road and inland waterway transport. However, we in Europe should agree that specific taxation of road transport in particular should be gauged by this yardstick and no other – so that road transit traffic is not subjected to any excessive financial burden. (Erdmenger:65)

Frohnmeyer (1983:10), another EEC official, is very sceptical about certain aspects of rail performance and sees roads as a better solution for local passenger traffic in rural areas: 'the road technique presenting inherent advantages to guarantee a sufficient supply of transport services.'

The scepticism about rail, and the acceptance of technical difficulties associated with social and environmental costs and benefits, has led the EEC to fudge the whole issue of infrastructure costs and their allocation and to prepare the way for a competitive environment which is still heavily biased in favour of roads. This conclusion has also been reached by the Conservation Society: 'It is a reasonable assumption that the majority and possibly the vast majority of investment in infrastructure would be spent on building new roads, and particularly motorways as part of the policies to liberalise road freight and to develop the regions' (Conservation Society nd. para. 3.12).

4 Railways

In spite of their decline compared with other modes, railways constitute still an important mode of transport in the EEC, and an area of conflict and debate at both national and EEC level about finances, investment, and closure proposals. At the EEC level there is a dislike of arrangements which support deficits in railway operation, and at the national level these deficits have led to closure proposals, as in the UK (Whitelegg 1984a), and massive new investments, as in France and Germany (TEST 1984).

Clearly the railway systems of twelve different European countries present a range of operational circumstances, finances, and investments, and in so doing, highlight one of the problems of the CTP – namely that of the attempt to impose a 'European logic' on the disparate and disunited transport systems of member states. In the case of railways, considerable political differences compounded by operational variations make any concept of harmonization considerably more difficult to implement than in the case of road transport. In the UK many rail systems have been closed over the past thirty years and new investment is shied away from. France and Germany have enthusiastically adopted new investment schemes in track and rolling stock for fast inter-city connections. The Portuguese provide a substantial degree of state support for their rail system and have the second lowest fares in Europe. Italians and Greeks are investing in new track, and in France and Germany controls on road haulage – which do not exist elsewhere – contribute to a healthier financial situation for the rail system than might otherwise be the case.

While space does not permit a detailed description or analysis of each rail system, it is nevertheless essential that some factual base is established to provide a background to assessment of the objectives of the CTP and its impact on European railways. TEST (1984) have compiled some valuable comparative statistics which make this task easier. Table 4.1 shows data for seven railways companies, five of which are members of

Table 4.1 Some characteristics of European railways and the countries in which they operate

		1972							1977							1981						
	Unit	BR	DB	FS	NS	SBB	SJ	SNCF	BR	DB	FS	NS	SBB	SJ	SNCF	BR	DB	FS	NS	SBB	SJ	SNCF
Population	million															56	62	57	14	6	8	54
Surface area	thousand km²															244	249	301	41	41	412	544
Population density	persons per km²															230	248	190	346	156	20	99
GDP per capita	ECU, current terms	2,573	3,820	2,015	3,135	n.a.	n.a.	3,416	3,918	7,357	3,342	7,065	8,313	8,639	6,331	7,998	9,995	5,554	8,894	13,250	12,169	9,542
Private cars	no./1000 population	233	260	229	218	244	303	269	264	326	290	278	306	346	315	283	285	339	324	370	348	349
National railway																						
Route-km	thousand km	18.6	29.2	16.2	2.8	2.9	11.2	36.5	18.0	28.5	16.3	2.9	2.9	11.2	34.4	17.4	28.4	16.4	3.0	2.9	11.2	34.4
Percentage electrified	%	17	32	49	58	99	62	26	21	37	52	61	99	62	28	21	39	54	61	99	64	30
Track-km	thousand km	48.3	66.8	29.4	6.8	6.7	17.5	n.a.	46.0	66.1	29.7	7.0	7.1	17.4	n.a.	42.8	65.5	30.4	7.1	7.2	17.1	74.0
Passenger-km, rail	billion (10)	28.3	38.8	35.4	8.0	8.3	4.4	43.0	29.3	37.3	38.4	8.0	8.0	5.4	51.6	30.7	41.8	40.1	9.2	9.1	6.9	55.4
Passenger-km, surface modes*	billion	416	470	341	106	64	c.76	346	452	616	405	127	c.75	c.97	n.a.	499	592	470	126	89	c.99	534
Passengers journeys, rail	million	754	979	355	184	224	57	614	702	971	394	171	205	65	661	718	1,103	396	205	218	79	687
Tonne-km, rail	billion	21.0	63.8	17.1	3.0	6.7	14.9	68.1	22.7	54.9	17.0	2.8	5.9	14.0	65.6	17.5	61.0	17.1	3.3	7.1	14.5	63.7
Tonne-km, surface modes*	billion	112	212	86	51	13	c.34	181	127	228	95	57	13	c.35	203	124	246	149	57	15	c.37	194
Freight lifted, rail	million tonnes	172	326	54	22	46	59	245	172	280	50	18	39	49	213	155	303	51	21	45	46	196
All rail staff	thousand	260	403	210	27	41	42	289	242	366	225	26	39	39	269	234	325	225	28	39	38	248
Railway operating staff	thousand	199	391	202	27	39	38	274	217	354	219	26	37	34	259	206	314	219	28	37	33	239
Public sector																						
compensation	million ECU								426	2,377	877	273	n.a.	105	1,531	686	2,790	1,194	259	n.a.	85	1,354
Compensation per track-km	thousand ECU								9.3	35.9	29.5	39.2	n.a.	6.1	n.a.	16.1	42.6	39.2	36.2	n.a.	5.0	18.3
Compensation per route-km	thousand ECU								23.7	83.3	53.9	95.9	n.a.	9.4	44.5	39.4	98.4	72.8	87.6	n.a.	7.7	39.4
Compensation per capita	ECU								7.6	38.7	15.5	19.6	n.a.	12.7	28.8	12.3	45.2	20.9	18.2	n.a.	10.3	25.2
Operating cost	million ECU								1,902	9,252	2,387	591	1,074	778	4,607	2,254	9,446	2,263	613	1,323	671	4,460
Income less compensation	million ECU								1,509	5,295	775	317	830	652	2,928	1,527	5,410	545	346	1,032	535	2,852
Other profits	million ECU								3.1		26.5				6.2		17.6					53.1
Loss	million ECU								389	3,957	1,585	273	244	126	1,673	717	4,035	1,700	266	291	136	1,555
Loss per capita	ECU								7.0	64.5	28.1	19.7	38.1	15.2	31.5	12.8	65.4	29.7	18.8	45.5	16.4	28.9

*1971, not 1972 data because there has been little change in population and surface area, these are not quoted for 1972 and 1977; financial data are missing for 1972 because Union Internationale des Chemins de Fer (UIC) data for that year are not compatible with their 1977 and 1981 data.

Sources: Eurostat (1983a); UIC (1973, 1979, 1983); Tanner (1983); Department of Transport (1979, 1982a).

the EEC, at three different time intervals.

In 1981 the French possessed the largest route-kilometres, with
Germany not far behind. There is wide variation in electrification, with
the UK lagging far behind and Switzerland having almost complete
electrification. The table shows how much work is done by the railways
in each country and how this compares with all surface modes for both
passengers and freight.

The financial status of railway companies is revealing about national
differences, and the loss per capita entry produces an interesting ranking.
Taking first the heaviest loss per capita, and then in increasing order of
financial viability, we have the following ranking: Germany, Switzerland,
Italy, France, Netherlands, Sweden, and the UK. This pattern reveals a
great deal about the status of railways in the respective countries, with
the 'best performer' (the UK) having the greatest difficulty in winning
approval for new investment and being the most eager to reduce track
length, services, and staff.

International comparisons of railway operation are notoriously
difficult to achieve, as Nash (1985) has pointed out: 'Readers should be
warned that the accuracy of such comparisons is always open to question

Table 4.2 Freight market share, 1970–84

	1970 (thousand million tonne-km)			1984 (thousand million tonne-km)		
	Rail	*Road*	*IW*	*Rail*	*Road*	*IW*
Germany	70.27	78.0	48.8	58.93	130.0	52.0
Belgium	7.8	13.0	6.7	7.9	19.8*	5.2
Denmark	1.8	7.8	–	1.6	8.1	–
France	70.4	66.3	14.1	60.1	88.4	8.8
Greece	0.7	6.9	–	0.7	8.0*	–
Ireland	0.5	–	–	0.6	5.0†	–
Italy	18.1	58.7	0.35	17.8	143.4*	0.27*
Netherlands	3.7	12.4	30.7	3.1	18.3	33.5
Portugal	0.78	–	–	1.24	11.8†	–
UK	24.5	85.0	2.0	12.7	106.9	2.4
Spain	10.3	51.7	–	11.9	111.5	–

*1983 data.
†1980 data.
Source: ECMT (1986a).

Table 4.3 Mean length of haul for freight, 1975 (in kilometres)

	Road	Rail	All modes
UK	60	119	77
West Germany	37	189	62[†]
France	58	292	100[‡]
Holland	47	154	58
Belgium	31	114	46
Italy	60[*]	345	94[§]

[*]1973.
[†]Excludes coastal shipping (0.2 per cent of tonnage).
[‡]Excludes coastal shipping (3.1 per cent of tonnage).
[§]Excludes inland waterways (0.3 per cent of tonnage).
Source: Nash (1985).

due to definitional problems (Nash: 238).

Nevertheless, Nash shows how a range of productivity and input-output measures can be compared successfully. From the point of view of transport policy at the European level, four characteristics are of particular interest: (1) the freight market share; (2) the passenger market share; (3) investment; and (4) financial performance. Table 4.2 shows the freight market share of the different modes in different countries.

Between 1970 and 1984 the work done by railways dropped in all countries except Portugal, Spain and Belgium. Germany's rail traffic fell 17 per cent, France's 15 per cent and the UK's almost half. Road haulage gains were much greater than rail losses: road freight in Germany, for example, rose by 60 per cent. In Italy, where rail freight has remained at approximately the same absolute level, road freight increased by 240 per cent. The UK boasted a 125 per cent increase. In 1984, rail's share of the freight market varied from 38 per cent in France to 10 per cent in the UK; Germany had 24 per cent and Italy 11 per cent. Different percentages would emerge if tonnes lifted were used instead of tonne-kilometres.

Several factors can be held to account for these variations. Much railway traffic is very commodity specific, and it has a large market share in heavy and bulky basic raw materials and goods, with a low share of certain manufactured commodities, e.g. food, drink, and tobacco. The railway market share will then be heavily influenced by the relative importance of different commodities in different countries. Length of haul is also important and something which is very much determined by geographical factors, such as the length of the network on the NS or EW

Table 4.4 Passenger market share, 1970–84 (in thousand million passenger-kilometres)

	1970			1984		
	Rail	*Bus*	*Car*	*Rail*	*Bus*	*Car*
Germany	37.3	58.3	350	39	70.5	490
Belgium	7.5	2.9	49	6.4	2.8	70
Denmark	3.3	4.6	33	4.4	8.1†	41
France	40.6	–	–	60	–	–
Greece	1.5	4.7	5	1.6	5.4†	11.2†
Ireland	0.7	–	–	0.9	–	
Italy	32.4	32.0	211	37.1	100.3†	334†
Netherlands	8	9.9	72	9	11.8	122.8
Portugal	3.5	4.3	17	5.4	8.3†	48.7†
UK	30.4	53	264	29.7	44	437
Spain	14.9	20.9	64	16.42	31.0	118

*1975 data.
†1983 data.
Source: ECMT (1986a).

axis and the population density, distribution of large cities and opportunities for international as well as national train movements. The possibilities for the UK and Eire to capitalize on international movements are obviously limited by geography, though the Channel Tunnel could bring about a major change in rail's share of freight in these countries when fully operational. In the case of French and German railway traffic, there is protection from road hauliers for long-distance traffic (over 150 km). This situation does not exist in the UK or Belgium where there is unfettered competition between road and rail, a situation the EEC would like to see become the rule and not the exception (Frohnmeyer 1983). Table 4.3 shows mean length of haul in kilometres for freight, allowing rail to be compared with other modes.

West Germany and the UK, with similar surface areas, perform very differently on this measure, where Germany has the benefit of longer-distance revenue-earning trains. France, with a much bigger surface area than any of the other countries shown in table 4.3, manages a mean length of haul of 292 km, only exceeded by Italy at 345 km. Geography plays a large part in these discrepancies at the mundane level of size and shape of the national territory, but also in terms of the relationship between domestic flows, import and export flows, and the importance of port-related activities. In the case of the UK, with a large number of ports and generally short distances from ports to manufacturing or distribution centres, it would be difficult for the length of haul to rise much above 100 km, and at these distances the lorry can be very competitive. The situation for passenger services is shown in table 4.4 for eleven countries and two time periods. Once again, it can be seen that there is evidence for the generally held view of rail's decline at the hands of private motoring. All the countries listed (where comparable data are available) have suffered a loss of rail passenger traffic.

This is part of the 'rail problem' (Pryke and Dodgson 1975) which has occupied so much time and energy on all sides of the debate surrounding closure, investment, rationalization, bus substitution, and rail track conversion to road. There is still no clear consensus on the basis for the discussions themselves. Does rail pay more than its 'share' of track costs in comparison with road? Does conventional cost–benefit analysis exaggerate the benefit of road schemes and underestimate the benefits of rail schemes? Why are proposals made for closing railways below certain critical thresholds of traffic or profitability, but not for closing roads on the same basis? Many of these issues are discussed in Whitelegg (1984a) and TEST (1986), going beyond the remit of a discussion about the CTP. Nevertheless, it needs to be understood that the whole background to railway economics and evaluating the performance of rail passenger services is inextricably involved with anomalous and discriminatory methods of supporting rail and road, and with highly politicized judgements about the need and value of different modes under different circumstances.

The nature of rail passenger travel is changing rapidly in Europe, the pace of change being much quicker in some countries than others. Reference has already been made to new lines and new services in West Germany, France, and Italy, serving longer-distance national markets. The nature of rail services in many urban areas is also changing rapidly with the construction of new and improved suburban services, underground lines, and new connections linking existing main lines as in Manchester, London, and Frankfurt. Hall and Hass-Klau (1985) have drawn attention to many of these developments in Germany and Britain, and the affects they are having on generating new traffic and contributing to environmental and economic improvements. These projects are discussed under the heading of urban rail investments later in this chapter.   RANSPOR

In rural areas also, there are signs that passengers can be won back to the railway system, as evidence on the Settle–Carlisle route in England, in 1986, has shown (Cumbria CC 1987). These are classic areas for line closure and service withdrawals, identified by the EEC as unsuited for rail operation (Frohnmeyer 1983). The problem with railways is not so much the standard economic or accountancy logic of the balance sheet, as expressed by Nash (1985:264), 'the real problem for rail passenger services has been a failure to achieve productivity at a sufficient pace to offset rising real staff, fuel and material costs', but more one of failure to adapt to new methods and new situations and to win political support for their undoubted energy, safety, and environmental and employment benefits.

The immediate manifestation of these failures is the loss of traffic, shown in table 4.4. The table inevitably glosses over some successes in winning back customers, as in France's TGV services and on new and improved urban systems. To some extent, the varying performance of rail, country by country, reflects the different strengths of rail's sub-markets: commuting trips to city centres, recreation trips, long-distance business trips, etc. Where central cores of cities have declined, as in London, this will lead to a loss of passengers that may not be made up from elsewhere, particularly if other factors such as bus deregulation (UK) are superimposed on an underlying downward trend. One of the few ways out of this dilemma is investment, and it is in the area of new investment in rail facilities that differences in Europe's railways really begin to emerge.

Investment

Figure 4.1 shows the geographical incidence of major investments in European railways and is based on the work of TEST (1984), which presents a thorough review of several rail strategy documents which identify plans for investment.

A plethora of institutions having an interest in railway planning in Europe has produced strategic plans for rail investment. Probably the most comprehensive of these is that produced by the Union Inter-nationale des Chemins de Fer (UIC): the Plan Directeur Européenne de l'Infrastructure (PDEI), initiated in 1970 and approved in 1973, and updated several times subsequently. The PDEI covered a network of 40,000 km and identified a number of axes of development:

- Basle–Milan via a new St Gothard Tunnel

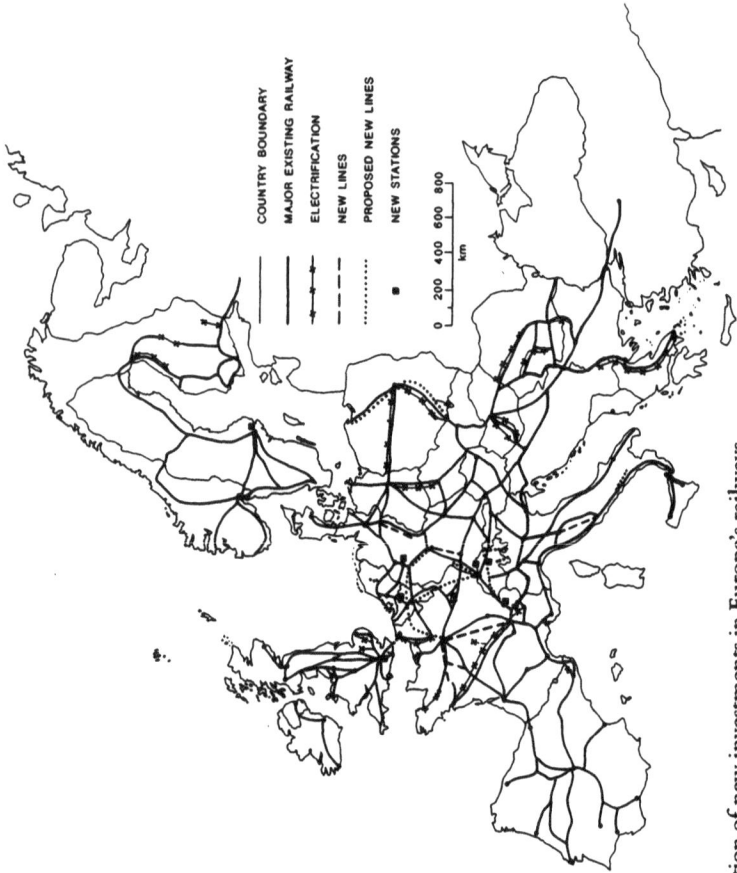

Figure 4.1 The location of new investments in Europe's railways

- Chambéry-Turin
- Munich-Innsbruck-Verona
- Barcelona-Narbonne
- GB-Continent via Channel Tunnel
- Budapest-Trieste
- Yugoslavia/Athinai to Sofia and Istanbul
- Vienna to Venice and Trieste
- Hamburg-Copenhagen-Stockholm/Oslo
- Lindau-Milan
- Munich-Salzburg-Ljubljana.

The identification of axes was supplemented by a great deal of detail on line improvement to permit high-speed running, electrification, signalling, and the identification of new stations. Much of this work is already completed or under way, particularly in those countries which have enthusiastically adopted rail investment. Other projects, such as the Channel Tunnel, have had false starts, while some links of obvious European importance, e.g. Munich to Istanbul, have received very little attention. In the UK, Spain, and Portugal there is a remarkable lack of investment activity, which leaves most of the investment which is going on in Europe in the hands of France, Germany, and Italy. The UIC is aware of these gaps and difficulties and has identified priorities for rail improvements:

- London-Paris
- Portugal and Spain to rest of Europe
- Budapest-Vienna
- Rhine/Ruhr-Basle-Milan.

Actual investments by representative European railways is shown in table 4.5.

British Rail's total investment has fallen by more than 71 per cent over the seven years, while that of Netherlands Railways has increased by 90 per cent. British Rail's steady and dramatic decline stands out in comparison with the sustained high levels of investment in Germany and France, which show only slight peaks and troughs. Investment levels in countries with much smaller populations and route-kilometres than that of the UK also identify the UK rail operation as outside the mainstream of European rail development.

In Germany rail investment has a 29.1 per cent share of all transport investment in the 1981–90 plan, compared with 16.4 per cent in 1971–80,

Table 4.5 Investment in European railways, 1976–82 (in million ECUs in real terms)

Year	BR	DB	FS	NS	SBB	SJ	SNCF
1976	683.57	1,155.79	–	106.17	–	–	616.77
1977	538.24	1,419.23	–	145.77	289.75	132.60	635.55
1978	515.30	1,434.96	–	–	288.36	118.60	–
1979	478.67	1,329.52	–	156.66	265.12	131.41	–
1980	342.55	1,306.47	–	173.49	249.00	172.26	672.64
1981	275.03	1,210.91	1,008.63	179.55	277.73	186.78	656.13
1982	201.92	1,196.76	–	201.93	297.23	189.41	594.28

Source: TEST (1984).

but 'what is more revealing is the change in emphasis as rail network expansion has a priority of all rail investment: 7.3% of the rail investment budget in 1971–80, 18% in 1976–85 and 38.5% in 1981–90' (TEST 1984: 35, vol. 1).

The scale of the German commitment to new track construction can be seen in figure 4.2. The reasons for this large-scale investment (see table 4.5) are a mixture of commercial and other factors peculiar to West Germany. With the partition of Germany at the end of the Second World War, the existing railway network was faced with very different demands to those which had fuelled its development. From being an east–west network with Berlin at its centre, Deutsches Bundesbahn (DB) now had to function on a north–south-oriented network with an emerging centre at Frankfurt. The north–south lines were old and of poor quality, and as Germany's post-war development accelerated, many became very congested. The two new lines under construction, i.e. Hannover–Würzburg and Mannheim–Stuttgart, are only the beginning of even more adventurous projects to link the Ruhr-Rhine/Main conurbations and improve links to Switzerland via the Heidelberg/Munich–Basle corridor (see table 4.6)

In 1985 the German government approved 35,000 million Deutschmarks for rail investment in the next ten years; 21,000 million Deutschmarks of this is specifically for new construction or improvement of lines, including marshalling lines and intermodal (road/rail) facilities. The high-speed rail network (200–250 km/h) which consisted

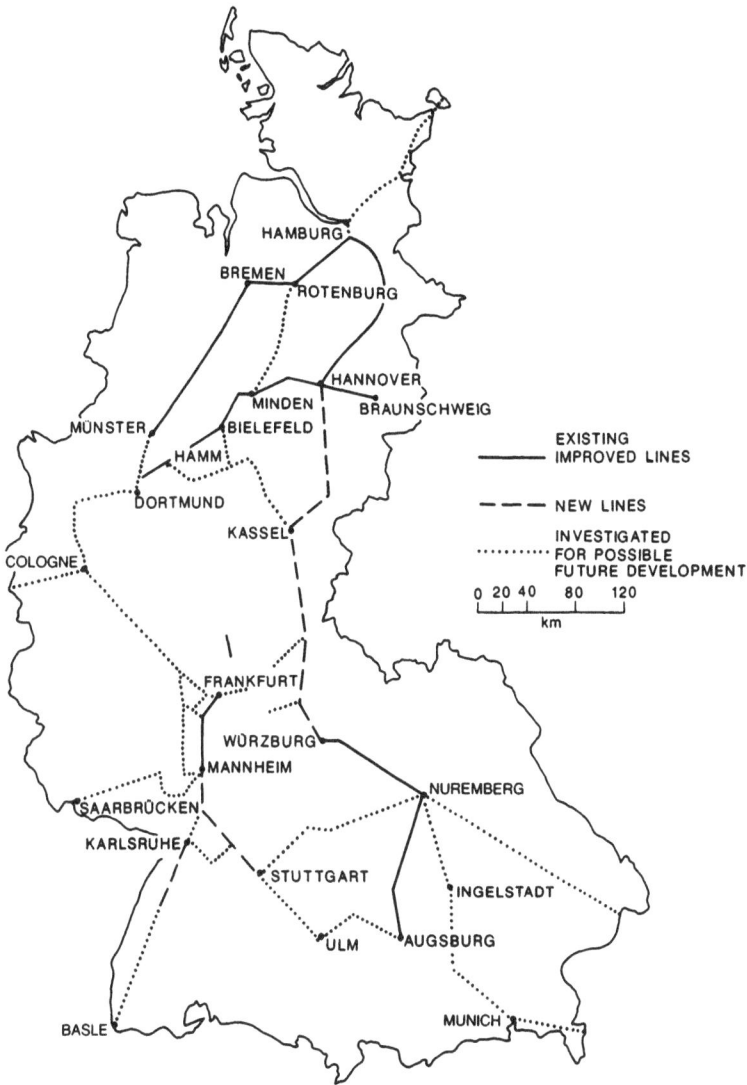

Figure 4.2 New lines in Germany

Table 4.6   New and improved lines of Deutsches Bundesbahn, 1985

New lines under construction		
Route	route length (km)	cost (million DM)
Hannover–Würzburg	327	11,000
Mannheim–Stuttgart	100	3,700
Line improvements		
Line	length of line (km)	length of high-speed section (km)
Hamburg–Bremen–Münster	287	186
Hamburg–Hannover	181	131
Dortmund–Hannover–Braunschweig	275	115
Giessen–Friedberg	32	–
Frankfurt–Mannheim	81	–
Würzburg–Nuremberg–Augsburg	240	37
Total	1,096	464

of 440 km in 1985 is planned to increase to 2,000 km by the turn of the century.

In Italy the Rome–Florence section of the Naples–Milan trunk route formed a bottleneck. It had low capacity and low speeds – and is being improved by the construction of new tracks to run in parallel. The Netherlands are constructing a new line to connect new settlements to the east of Amsterdam, and another from Amsterdam to Schipol Airport towards Den Haag. Switzerland proposes a new line from Lausanne via Berne to Basle and Zurich and there are also proposals for Alpine tunnels.

The construction of new lines in Germany is not without opponents (Holzapfel nd). There are environmental arguments against new lines which should not be overlooked just because the mode of transport is rail – not the usual object of opposition, roads. New lines tend to have few stops *en route*. The number of stops is crucial in the drive for greater speed and faster journey times, and several stops could eliminate a large percentage of the savings in journey time from the new tracks and higher-speed running. Fewer stops means a poorer service to inter-mediate towns and communities and a centralization of rail activity which may not be in the best interests of rail travellers. Another problem is the relationship between new lines and their investment requirements, and the degree of commitment to the remainder of the systems. There is more than a suggestion that commitment to new lines in Germany also involves closure of the less glamorous rural and lower trafficked routes. It is too early to be conclusive about this, but the emergence of two very different markets for rail travel would have serious effects on the 'second division' railways.

France has more experience with its new high-speed lines than has Germany and its Trains à Grande Vitesse (TGV) are much quoted as a model for this kind of development. Their place in the French railway network can be seen in figure 4.3, and there is a full discussion in TEST (1984), on which the following account is based.

The TGV was first proposed in 1969 and the Paris–Lyons corridor was selected as the route for construction of new track, mainly to relieve congestion at existing bottlenecks. The main elements of the TGV are new track, constructed to high standards of alignment (but with steep gradients), a shorter route as a result of the straightening, new and powerful locomotives and standards of speed and comfort, and a marketing image which will attract passengers and provide effective competition to other modes of transport. Final approval for the project was given in 1975; construction began in 1976; and the southern section opened for service in September 1983. By 1984 the trains were running through to Switzerland, St Etienne, Chambéry, Besançon, and Marseilles, as well as the original destination, Lyons.

The TGV project represents a very large investment in rail transport. At June 1982 prices, the cost of building the new line is put at 7,004 million francs. Rolling stock consisting of eighty-seven train sets (each with eight passenger cars and a power car at each end) cost 4,570 million francs, a unit cost of 47.6 million francs. Expenditure on TGV–South-East over the formative years of the project (1975–1983) is shown in table 4.7.

The TGV is proving to be a profitable investment in spite of heavy loan charges. Immediate returns on the first year of operation were 16.4 per cent. In the analysis conducted by TEST (1984) it was concluded that the balance between takings and charges on all TGV traffic came out in favour of the line. The first year of operation produced revenue of 1,039 million francs, against operating expenditure of 504 million francs, and this relates to only one-third of the system being operational. The success of the TGV in attracting passengers is outstanding; table 4.8 shows the results of the first eighteen months. The passengers have been won from conventional rail, as well as air and car travellers. The proportions were

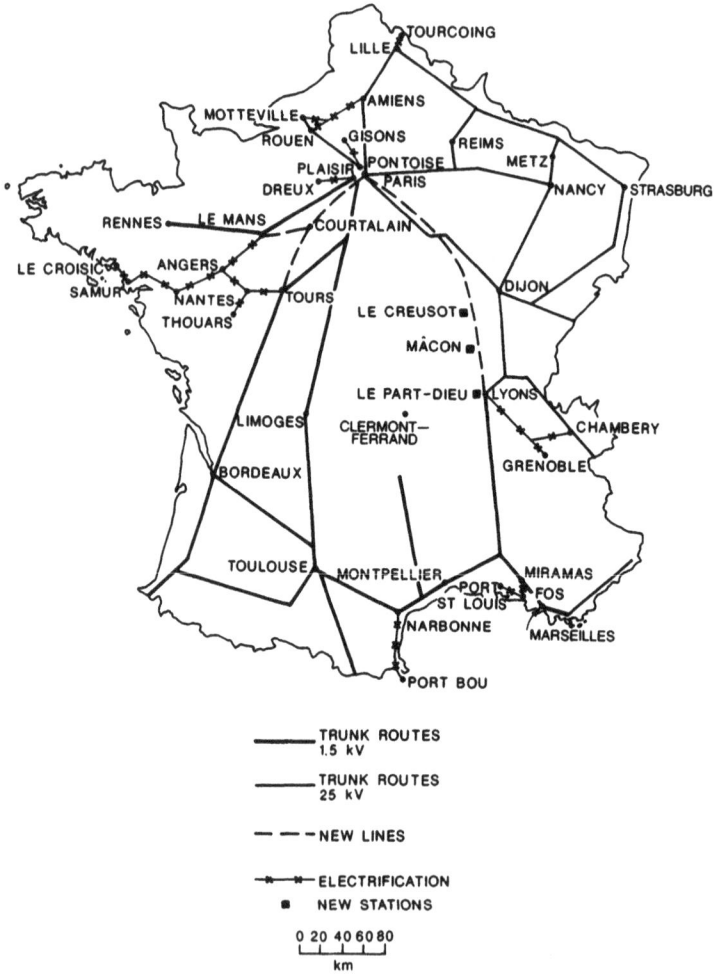

Figure 4.3 Major investments by the French railways (SNCF)

Table 4.7 Expenditure on TGV-South-East, 1975–83 (in million francs, at June 1982 prices)

	1975	1976	1977	1978	1979	1980	1981	1982	1983+	Total
Construction	20	51	321	1,060	1,332	1,582	1,413	825	400	7,004
Rolling stock	45	45	196	103	483	1,266	889	1,054	534	4,615
Total	65	96	517	1,163	1,815	2,848	2,302	1,879	934	11,619

12,600/17,800 from conventional rail, 1,100/17,800 from air and 'the occupants of 200–800 cars' (TEST 1984: 72, vol. 2).

A study of the effects of German high-speed rail routes has predicted a 15–17 per cent loss of air traffic in favour of the new rail lines (Haupt and Wilken 1985). Clearly high-speed services have the capacity to make considerable dents in the market share of domestic air services, an effect which is likely to become more pronounced on a European scale as investment projects, particularly the Channel Tunnel, come to fruition and deregulation in the air gathers momentum.

The TGV experience gained on the Paris–Lyons project has been extended to south-west France in the TGV-Atlantique, with a decision to proceed in late 1983. The TGV-Atlantique involves concentrating mainline traffic of the Paris–Le Mans and Paris–Tours trunk routes on to a new high-speed line, greatly improving rail services in western and south-western France. Both branches of the line (see figure 4.3) are 308 km in length, and the construction costs of line and terminals was 7,600 million francs at 1982 prices. Ninety-five TGV train sets are required for these new services, running at 220 km/h and costing 4,400 million francs at 1982 prices.

The estimated benefits and financial performance of this project are both healthy. The internal rate of return on the investment was 12.9 per

Table 4.8 Passengers carried on TGV services during first eighteen months (thousands)

Route	1st class	2nd class	Total
Paris–Lyons–St Etienne	1,787	4,323	6,110
Paris–Midi	410	1,451	1,861
Paris–Geneva–Chambéry	285	882	1,167
Paris–Dijon–Besançon	176	462	638

cent, and for the community (i.e. including social benefits and environmental improvements), 23.6 per cent. A financial analysis showed
recouping of the total debt within ten years. In terms of the market
performance of the new lines, out of a total of 21.4 million passengers
they are expected to win 16.1 million passengers from conventional
trains, 0.8 million from air, and 4.5 million from road transport. The
remainder will be newly generated traffic.

Other TGV projects are planned, or have existed for some time and
been shelved because of uncertainty over the Channel Tunnel. Among
these is the Paris-Brussels-Cologne line and an extension to London via
the Channel Tunnel (Kracke *et al.* 1986).

The French national railway, SNCF, has not limited itself to the TGV
in investment plans and is also involved in electrification (e.g. 230 km
from Le Mans to Angers and Nantes) and the improvement of cross-
country services. The TEST report (1984) estimates that the cost of
electrification programmes – under way or completed – amounts to
36,425 million francs.

Freight transport by rail in France is on a much larger scale than in
many other European countries and this is, in part, due to the large
investment in private rail sidings. In France there are over 10,000 private
rail sidings, compared to over 15,000 in Germany and approximately
1,500 in the UK, and 90 per cent of freight traffic originates, terminates,
or does both, at these locations. However, the French situation is
somewhat more complicated by its restrictions on goods vehicle licensing
and by its toll motorways. Both factors undoubtedly contribute to
France's 72 billion tonne-kilometres of freight by rail, compared with 98
billion by road (1980 data).

The French and German investment experience is echoed in the
Italian state railway system, FS. Here a new line is under construction
from Rome to Florence, a total length of 252 km, and at a cost of 3,003
billion lire (2.27 billion ECU). The FS have other large-scale investment
plans which include electrification of lines, modernization of rolling
stock, and improvements to freight. The investment levels are high
compared with other European countries – 3,284 billion lire in 1981,
3,507 billion lire in 1982.

Britain's record of investment does not stand comparison with the
three countries briefly outlined so far, and indeed lags behind many
countries with much smaller populations (table 4.5). It would be a major
task of political rather than economic analysis to explain this (Whitelegg
1987a), but the facts themselves are indisputable. The long-drawn-out

debate over electrification of the East Coast Main Line (ECML) is symptomatic of the malaise. Only started in 1985 and costing an estimated £242 million from London to Newcastle upon Tyne, the scale of the project is much less than in continental electrification schemes and shows a much better rate of return than the Department of Transport's test discount rate (t.d.r.) (an average of 12 per cent compared with a t.d.r. of 7 per cent).

TEST (1984) makes the point that British Rail's investment proposals are little more than renewal in character, and are not innovative in the sense of West German or French schemes for winning new passengers on new tracks. They are also financed entirely by the railways themselves, not by the state. TEST goes on to identify a series of possibilities for rail

Table 4.9 TEST proposals for rail investment in the UK

Rail access to airports
Heathrow: (1) Heathrow Parkway on the M25 possibly at Iver on the line to Reading; (2) a direct BR line via a 9-km tunnel from Iver to Feltham via Heathrow central

Gatwick: electrification of Tonbridge–Redhill and Reigate–Guildford and Wokingham.

Manchester: loop to airport from main London line at a cost of £38 million

Birmingham: improved services from Birmingham International as far afield as Leicester

Edinburgh: a new station on Edinburgh–Aberdeen line to serve the airport

Rail market opportunities
Exploiting corridors of movement where Department of Transport road schemes frequently identify demand for travel, e.g. Birmingham–Nottingham, South Coast links (electrify Ashford–Hastings), north of Inverness rail links (including the now rejected Dornoch Firth rail bridge), station re-opening

Cross-city links
London: double-track deep tunnel from Battersea to north of Euston; Blackfriars–Farringdon disused line
Manchester: Windsor curve and Hazel Grove Cord are now under construction linking Manchester's two railway systems

investment in Britain which would increase market share for both freight and passenger and make a noticeable difference to BR's much-quoted weak financial situation (although it did declare a profit of £1 million at the end of the 1985/86 financial year). The main categories of new investment are summarized in table 4.9 excluding the Channel Tunnel, which is dealt with in more detail in chapter 10.

Urban rail investments

An important dimension of rail networks in Europe is their role in cities. Railways carry out important functions of distributing workers in conurbations and of supporting conurbations as retailing and service centres. This dimension of rail activity does not seem to have figured at all in CTP thinking, despite the vitally significant role of railways in

Table 4.10 Some rail system investments in European cities

Barcelona	New loop line to take freight out of city tunnels; change from bus routes alongside rail to ones feeding rail stations
Hamburg	Extend S–Bahn to Neugraben, and add further link tracks near Hauptbahnhof
Copenhagen	Likely new north–south S–Bahn type in 1990s
Lille	Plan to add three further metro lines to that operating; Upgrade SNCF services, co-ordinate fares
Madrid	Rail plan envisages increased share suburban transport from 10 to 25–30 per cent; cross-town Renfe tunnel connecting with most metro lines; new tunnel connecting main stations
Milan	Separate freight from passenger, introduce new rolling stock and peak-hour trains, reorganize bus services as Barcelona, connect Garibaldi and Vittoria terminals across city centre
Munich	Highly comprehensive DB, U– and S–Bahn rail network to be extended further, remaining bottlenecks removed, S–Bahn line 3 extended 19 km to new airport
Naples	Track doubling, new rolling stock and signalling, modernize tramway and three funiculars, construct first metro line
Vienna	Improve and expand existing S–Bahn network and its interchange with U–Bahn

Source: TEST (1984).

urban areas and of the heavy drain on rail finances of supporting very large peak-hour demands in cities.

Recent studies (TEST 1984; Hall and Hass-Klau 1985; Roberts 1987) show a large number of urban investment schemes prompted by the

Table 4.11 Existing and planned U-Bahn systems and light rail systems in Federal Republic of Germany, including West Berlin

Town	System	Length[a] (km)			First line
		Existing	*Under con-struction*	*Planned*	
Berlin	UB	101	8.2	200*	1902
Hamburg	UB	89.5	6.5	120	1912
Munich	UB	37.5	14.5	90	1972
Nuremberg	UB	15.0	4.8	42.1	1972
Bielefeld	LR	5.1	3	41	1971
Bremen	LR	10.0	–	39	1968
Frankfurt	LR	41.3	10.5	50.8	1968
Hannover	LR	60.7[a]	6.5	98	1975
Cologne	LR	41.6	5.0	continuation of changing the tram to LR	1968
Bonn	LR	25.5	3.1	?	1975
Ludwigshafen	LR	9.0	2.8	23.4	1969
Mannheim	LR	1.8	–	?	1971
Stuttgart	LR	39.5	4.0	100	1962
Wuppertal	PR[b]	13.3	–	0	1901[b]
Stadtbahn-Rhine-Ruhr[†]		53.9	42.5	139	1977

*For whole of Berlin.
[†]Light rail system includes the cities of:

Bochum	3.3 km
Dortmund	13.0 km
Dusseldorf	1.6 km
Essen	10.1 km
Gelsenkirchen	0 km
Herne	0 km
Mülheim/R	5.5 km.

[a]Track length.
[b]Schwebebahn (Monorail).
Source: Hall and Hass-Klau (1985).

necessity to deal with congestion and other car-related problems, as well as the need to invigorate and revitalize cities. Table 4.10 details some of these schemes.

The German expenditure both of effort and cash on such projects is remarkable and contrasts markedly with the neglect of urban rail transit in British cities. Many medium-sized German cities (150,000 population) have some form of train system and larger cities, such as Hamburg, Nuremberg, Frankfurt, Berlin, Munich, and Cologne, have underground systems as well as S-Bahn (commuter rail system) and a tram system.

In Munich the public transportation system had a total length of 543 km and a route length of 2900 km (1981). In 1981 the total U-Bahn system had a route length of 40.2 km, with new extensions opened in 1983. Munich is an exceptional investment which was boosted, in 1972,

Table 4.12 Operating results, 1983

Network	Billions of national currency units		
	Operating costs (A)	Revenue from traffic (B)	B/A × 100
Austria	26.8	14.3	53.3
Belgium	100.1	25.7	25.6
Denmark	6.8	3.3	48.5
Finland	3.4	2.2	64.7
France	66.1	34.4	52.0
Germany	30.4	17.0	55.9
Greece	15.2	6.0	39.4
Ireland	0.296	0.191	64.5
Italy*	10.7	2.2	20.5
Luxembourg	7.5	1.6	21.3
Netherlands	2.75	1.35	49.0
Norway	3.521	2.243	63.6
Portugal	26.7	11.2	41.9
Spain†	210.0	119.6	56.9
Sweden	8.302	6.549	78.8
Switzerland	4.043	2.864	70.8
Turkey	98.1	32.9	33.5
UK	2.968	1.914	64.5

*In thousand billions.
†Ministry of Transport.
Source: ECMT (1986a).

by the Olympic Games. Other German cities are impressive still, described in some detail in Hall and Hass-Klau (1985). Table 4.11 summarizes the situation as far as a selection of German cities is concerned.

In the UK (outside London) the only comparable system is the Newcastle-upon-Tyne metro, completed in 1984, though Liverpool and Glasgow have had some modest investment. Manchester is probably the largest city in Europe without any light rail transit or underground system and the city is further disadvantaged by its legacy of two central railway stations, 1.5 km apart, connected only by an inconvenient shuttle bus service.

The London rail system is very heavily utilized and in need of new investment (GLC 1986a). The 326 BR stations are used each day by 200,000 London commuters, whereas the 225 underground stations are used by 250,000 commuters. The two systems (BR and London Regional Transport which operate the buses and underground trains) are not well integrated and offer considerable scope for service improvements. The GLC report describes the importance of rail investment to the urban economy and its importance in meeting the transport needs of women, ethnic minorities, and all groups in search of better work opportunities. None of these issues figures in CTP statements on rail transit.

The suggestions by TEST detailed in table 4.9 are not unusual by European standards. Many cities have much better through-rail services than exist in Britain (e.g. Paris, Brussels, Madrid) and others are under additional consideration for improvement (e.g. Frankfurt, which already has much better transfer facilities than any London terminus, is considering a new tunnel to link Frankfurt Hauptbahnhof with OstBahnhof).

Clearly investment is something which does discriminate between European railway operations and the respective transport policies of European nations. It is, however, a means to an end, and in the highly competitive world of transport that end is viewed through the balance sheet. In the next section we turn to railway revenues and railway finances.

Railway finances

Table 4.1 shows the basic financial data relating to a selection of EEC and non-EEC countries; in table 4.12 this is updated and enlarged to all twelve EEC member states and other European countries (UIC 1985).

Table 4.12 shows much diversity in the financial status of railway operations in Europe. Railways of EEC states fall into four groups in terms of the proportion of their operating costs which they meet from revenue. The groups are (in descending order of importance of revenue):

group 1 (60 per cent)	Ireland UK
group 2 (50–60 per cent)	Spain Germany France
group 3 (40–50 per cent)	Netherlands Denmark Greece Portugal
group 4 (30 per cent)	Belgium Luxembourg Italy

RANSPOR

Ireland and the UK have to meet a much larger proportion of their operating costs from revenue than do the other countries. Italy is the country with the lowest revenue contribution, at 20.5 per cent. Outside the EEC, there are examples of railway operations which contribute even more than the UK from their revenue; Sweden contributes 78 per cent and Switzerland 70.8 per cent.

Measuring the financial performance of railways is fraught with difficulties even within one national system, where accountancy conventions and relations with government over subsidies can make apparently simple concepts like profit and loss treacherous indeed. If this is a problem at the national level, it is all the more intractable at the European level.

Nash expresses some of these difficulties:

> First, most Western European railways receive explicit subsidies for particular obligations regarding fares and service levels. In principle these represent the cost of departing from mainly commercial policies in the particular respect in question, but given that most of the

European rail network in commercial and social traffic alike would disappear in the absence of subsidies such an explicit interpretation is often difficult. Second, most European railways have still been declining for many years, a loss after subsidy, which governments cover by grants or loans. (Nash 1985: 265)

Nash goes on to explain that many other factors may distort the true picture. For this reason, he chooses to measure financial performance as a single ratio of traffic receipts (excluding subsidies) to current and capital expenditure for the year in question, except for any expenditure relating

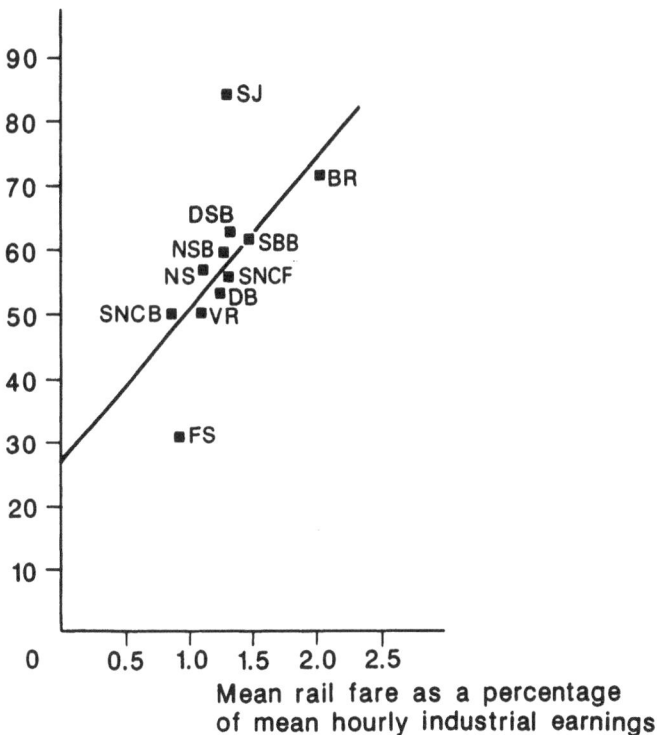

Figure 4.4 The financial performance of railways

to wholly new lines. Nash's results are reproduced in Figure 4.4.

Nash shows a very clear relationship between financial performance on this measure) and fares relative to wages. The diagram shows that high-fare railways (in 1977) received the smallest subsidy. This is still the case in the UK, though Sweden's fare policy has now altered dramatically with lower fares being used as a way of winning a greater market share. Figure 4.4 clearly identifies three groups of railway operation in terms of degree of subsidy. Group one, with the lowest subsidy and the highest percentage of current and capital expenditure covered by revenue, contains Sweden and the UK. The second group, which is not clearly differentiated internally, consists of Denmark, Switzerland, Norway, France, Germany, Belgium, and Finland, and lies in the range 50–65 per cent in terms of revenue contribution to current and capital expenditure. Italy is the only representative of the third group and is clearly the most heavily subsidized, with revenue contributing only 30 per cent of current and capital expenditure.

The information in TEST (1984), reproduced as table 4.1, produces slightly different rankings. Taking compensation per route-kilometre, Sveriges Statens Järnvagär (SJ) and BR are still the most heavily subsidized, but Société Nationale Chemin de Fer (SNCF) and DB change places, with DB receiving more subsidy than its French counterpart. The obvious point to be made is that there is no absolute measure of financial performance, but there are considerable variations within Europe in attitudes to subsidy, and these pose real difficulties for an emerging Common Transport Policy. Areas of particular difficulty for an emerging CTP are the ground rules for pricing rail services, standardizing these methods, agreeing rules for the allocation of subsidy, and bringing rail and road subsidies into a common framework of evaluation. We must now turn to the CTP itself and examine its attitude to rail operations and attempts to intervene in this particular transport market.

Railways and the CTP

The general orientation and background of the CTP have already been discussed. The Commission's emphasis on fair competition between modes has led to specific rail regulations on a public service obligation (PSO) subsidy for unprofitable but socially necessary services, normaliz-ation of accounts, and harmonization of the relations between railways and states.

According to Erdmenger (1983), there were two prime aims for action in the area of railways: the elimination of distortions of competition which arose, to the detriment of railways, from the traditional state intervention in this sector; and the attainment of better transparency of state financial contributions to railway undertakings.

The decisions which advanced these aims were Regulation 1191/69 on the elimination of, and possible compensation for, public service obligations; Regulation 1192/69 on the normalization of accounts of railway undertakings; and Regulation 1107/70 on aids to inland transport modes. The substance of these three Regulations is summarized in table 4.13; these have the force of law and the EEC Commission has powers to see that they are obeyed. The Regulations were incorporated in a later measure (a Council decision of May 1975) which attempted to provide greater autonomy to railway administrations of member states.

Independence should include separation of assets, budgets, and

Table 4.13 EEC railway regulations

Regulation 1191/69 on action by member states concerning the obligations inherent in the concept of a public service in transport by rail, road, and inland waterway; this Regulation applies to all three modes and seeks to reduce to a minimum the social grants or obligations, which benefit a mode and hence cause distortion in the conditions of competition; its aim is to 'terminate the public service obligations defined in this Regulation', but there is acknowledgement of the need to ensure the provision of adequate transport services. 'However this Regulation strikes at the very root of the concept of providing public transport as a social service. Article 3 states that where it has been decided to maintain a PSO, and where it can be done in more than one way while satisfying similar conditions: "the competent authorities shall select the way least costly to the community"' (Conservation Society, nd)

Regulation 1192/69 on common rules for the normalization of the accounts of railway undertakings; its purpose is to provide for the payment of compensation by the member governments to their railway undertakings; the imposition of unfair financial burdens and the granting of benefits to the railways that would be liable to cause substantial distortion in the conditions of competition must be eliminated.

Regulation 1107/70 seeks to limit the circumstances and cases where aid may be granted to certain kinds of co-ordination of transport, research and development and social obligations as under Regulation 1191/69.

accounts from those of the state and 'sufficient independence as regards management, administration and central control over administration, economic and accounting matters with a view to achieving financial balance'. The basic position of the EEC Commission was, according to Bayliss (1979:37), 'that it would, as a long term objective, like to see railway undertakings which are managed on normal business principles, and which are independent of state financial support'.

There is no recognition within these early EEC regulations and pronouncements on railways of their wider social, economic, and environmental role and of the possibilities for investment and expansion to produce widespread European benefits. TEST draws attention to this lop-sided view of railways in the Community:

> [The policies] are very limited and say nothing about railways as savers of human life, time, energy, natural resources, and congestion ... Comparative levels of subsidy between rail and road are never questioned. (TEST 1984: 27, vol. 1)

Gwilliam (1980) takes the view that the regulations themselves have not been effective. Basically they were intended to limit the level of subsidy required to those goods defined in the Regulations, and subsidies resulting from other causes were to be eliminated:

> But in fact the vagueness of the limiting regulations and the immense difficulties of interpreting railway costs have meant that the regulations have not been effective. In most cases deficits are allocated between the regulations ex-post to exhaust the total deficit emerging rather than ex-ante as a management guide. (Gwilliam 1980: 51)

The situation of the railways has, therefore, been little changed by these EEC interventions. Erdmenger (1983) laments the growth of state aid to railways (an increase of 60 per cent between 1973 and 1977) and a total charge of 12,000 million ECU. The consequence for EEC thinking has been a shift of emphasis away from direct attempts to limit subsidy and a move towards increasing efficiency either by exhortation or by actually doing something to improve infrastructure. We have already seen in chapter 3 on infrastructure how EEC thinking has developed in this area and how rail projects have figured in the earliest projects to be selected.

Erdmenger (1983) identifies a further area for improvement, that is the promotion of international combined transport utilizing both rail and road in co-operation rather than in competition. This would indeed be progress, but there is little to show for its identification either in concrete reality or in regulations and funding (this theme is taken up in a later chapter).

An important development in EEC pronouncements on railways is continued in a Communication from the Commission to Council in February 1983 (EEC 1983b). Strictly speaking, this concerns infrastructure costing and charging:

> the railways should be put on the same footing as the two other inland transport modes ... the cost of providing and maintaining the rail infrastructure should, as with roads and inland waterways, be the financial responsibility of the state. (p. 21)

This is superficially an attractive solution to railway financial problems. It has certainly been advocated over the years by railway support organizations and it would, at least, allow the true level of state support for all transport modes to be revealed in stark detail. It is attractive at the EEC level because it brings about circumstances more conducive to normalization of accounts and the termination of distortions and irregularities which impede the competition between markets. What it would also do, of course, is give the state complete power of investment and disinvestment over a rail system in a way which might well conflict with the operator's view of market opportunities and with the users' view of the need for particular services. In Britain, at least, it could unleash an avalanche of closures for which Dodgson (1984) and Serpell (1983) have provided some partial justification and the Transport Users' Consultative Committee's procedures have done something to restrain. Should the EEC be successful in promoting this change in the relationships between state and railway company, then we might indeed see some radical changes in the size of railway networks, the markets they pursue, and the size of public subsidy they consume.

Frohnmeyer (1983) pursues this theme but, in the same address to a London conference, advocated the closure of rural railway lines and bus substitution, and suggests that there is 'scarcely a case for continuing [diffuse wagon load freight] the road technique presenting inherent advantages to guarantee a sufficient supply of transport services' (p. 10). It is interesting to note that the Commission justifies these proposals for reforming rail finances in terms of saving expenditure under national budgets. As this is not properly a function of the EEC and does involve value judgements about priorities which are determined nationally, one can assume that it is a general rationalization of the urge to do something about the 'rail problem'. It is disappointing that nowhere in EEC analyses is the 'rail problem' adequately defined and objectives for this transport sector clearly set down in any form other than financial targets.

The situation of railways remains relatively untouched by the activities of the CTP. The scale of subsidy is still larger than the EEC would prefer, the pursuit of national goals and investment strategies is still very much outside a European framework, and there has been very little progress in getting states to agree on any form of equal treatment of rail and road. This is the case in those countries which give rail encouragement through controls on road haulage over certain distances (e.g. Germany), and in those countries (e.g. Britain and Belgium) where the road sector is dominant in freight markets. The infrastructure policy of the EEC, and attempts to foster co-operation and European thinking among railway companies, may yet produce results in the form of a greater market share for rail freight, but the EEC's fundamental ambivalence over rail transport manifests itself as expressions of support at the same time as concepts of efficiency and least-cost forms of transport are promulgated as EEC goals. The latter do not do justice to rail's role and potential in inter-modal transport and take no cognisance of rail's environmental, safety, and economic benefits to rural and urban areas alike. The CTP has failed to appreciate the importance of railways as a mover of freight and people, and failed to give a lead for member states who, struggling with financial problems, were (and still are) receptive to positive transport policies as opposed to financial imperatives dressed up as transport policies.

5 Road transport

In one of its public relations pamphlets the EEC has spelt out what it regards as important or noteworthy in its approach to road transport (EEC 1983b). The major concerns are:

1. Road traffic accidents and the European Road Safety Year (1986)
2. Harmonized standards for brakes, lighting, windscreens, sound levels, etc. to permit free and safe movement of cars from one country to another; common standards on weights and measures of commercial vehicles (finally agreed at the end of 1984); and progress on lead-free petrol and its availability and controls on car exhaust emissions
3. Delays at frontier crossings and measures to reduce or eliminate them; allowances for fuel in tanks of commercial vehicles, single customs document, European passports, and driving licences
4. Making conditions easier for international road haulage by introducing community quotas which give more freedom to individual hauliers: 'In the long run the Commission hopes to liberalise the road transport market completely'
5. Social legislation covering standards for training of drivers, driving periods and rest periods, and tachograph regulations
6. Pricing systems and 'improved' taxation systems for commercial vehicles; the problem of transit routes and taxes through Austria and Switzerland

These concerns, with the exception of the first, all come firmly within the liberalization ideology of EEC intervention which has been discussed in chapter 2. They all relate, in one way or another, to the desire to improve competition between modes, reduce or eliminate the distorting influence of state action, and lower the costs of transport. These measures will be most efficacious, according to EEC policy, in bringing about

European integration and better prospects for economic growth. Erdmenger (1983) puts it more bluntly: 'The Common Transport Policy has made every effort to free the international carriage of goods in the Common Market from state intervention.'

State intervention is perceived as damaging to transport systems or, at least, as unnecessary, and in this lies one of the greatest paradoxes of the whole CTP effort, a body (the EEC) which regards most, if not all, national bureaucratic intervention in transport as irksome has taken on the role of a supra-national dispenser of regulation and controls. In so doing, it tries to tackle almost every aspect of transport operation (including social and environmental) but is locked in a framework which will encourage more traffic and more road haulage and, therefore, more environmental difficulties and social problems which will need a solution. The CTP, as it is presently constituted, cannot break out of this circle and its actions can only be interpreted by reference to an explanation which is based on its internal contradictions. The road traffic accident concern falls outside this framework and is not, in any case, a major CTP or even an EEC initiative; this theme is revisited in a later chapter.

It is not surprising, therefore, that the EEC's 1985 checklist does not reflect a particularly well-thought-out view of road transport and its place in a wider transport policy. This is not the place to set out in any detail what such a wide transport policy might look like. Both Adams (1981) and Plowden (1985) give a more than adequate view of this general level of argument. A little digression is, however, unavoidable.

Road transport, more than rail transport, must be viewed clearly from some general transport perspective. Its infrastructure demands are much heavier than rail; it kills more people than rail; its impact in urban and rural environments is much more dramatic than rail; and its provisions at the expense of the state (generally), and use by many different classes of consumer, makes its impact less generally predictable, but still intense.

Road transport offers complete personal control of movement for those able to afford the vehicles and confers economic advantages, as a result, in terms of access to jobs, shopping, and leisure and recreation facilities. In the case of road goods transport, there is a clear parallel within the firm and its control and organization of its transport fleet. There can be little doubt that the combination of personal control, state finance of a dense network, and the ability to respond to changes in the location of market opportunities has created a technological dominance for private motorized transport that is well entrenched in contemporary

European societies. It is a technology with many harmful consequences (of which road traffic accidents are only one manifestation) and a technology which, until recently, was most successful at eradicating its opposition, i.e. public transport. As a result of these constraints, any national transport policy will find its operating environment difficult indeed, and this is why transport policies at national level are rarely spelt out with clear objectives. One should not be too critical, therefore, if the same situation has evolved within the CTP. At both national and EEC levels a distinction can be made between policies which stimulate movement as a goal and those which are concerned with meeting basic needs and providing the right circumstances for balanced growth and environmental protection. In the former case, greater and greater levels of car ownership, passenger-kilometres or tonne-kilometres, are taken as signs of progress or, at least, of demand which must be met as far as financial resources permit. In the latter case, movement is not seen as a desirable goal and indeed may be seen as a strong negative consequence of misplaced investment in transport infrastructure. In the case of passenger movement, road space will be necessary for personal vehicles, for collective transport such as buses and minibuses, and for cyclists and pedestrians, though there is a strong case for specially designated, segregated facilities as in the case of pavement and pedestrian zones. The provision of more road space in response to rising levels of car ownership, demands fuelled by vested interests, or attempts to reduce congestion, may be counterproductive in the extreme and will almost certainly injure the interests of non-motorized road users and local residents. A transport policy for road passenger transport (as opposed to one for cars or lorries) will be mindful of these consequences and will consider accessibility to a wide range of facilities for a wide range of consumers, as more important than mobility, speed, and convenience for the car owner. In the cases of road freight transport, the considerations necessary for a balanced transport policy would not be locked into the unproven but strongly held belief that more road space equals a better economy. Rail and road can and do function well together, given the investment and the back up policies in land use and planning. Investment in rail sidings for private industry and protection for local residents blighted by lorry traffic are legitimate instruments of national transport policies and examples of a balanced approach which is lacking in the EEC checklist.

The remainder of this chapter will be concerned with the intervention of the CTP in road transport and hence will follow the development of those items in the checklist. This will then broaden out to consider the

Table 5.1 Major CTP initiatives in road transport, with reference to original sources

---

Proposal (OJ L 16, 1979) for harmonizing national law regarding a weight limit of 44 tonnes for five-axled articulated or semi-articulated lorries

Community permits to enable a certain number of long-distance hauliers to carry goods between all member countries without the need to obtain national authorizations (OJ L 358, 1977; OJ L 366, 1978)

Specification of maximum permitted driving hours per day and per week (OJ L 334, 1977), and requiring the installation of the tachograph in heavy road vehicles (OJ L 181, 1973)

Minimum levels of training for drivers (OJ L 357, 1976)

Mutual recognition of national certificates and qualifications, so that hauliers can set up businesses in more than one country (OJ L 334, 1977)

Community-wide driving licences and acceptance of national licences throughout all member states

Fixing of rates for carriage of goods by road between member states (OJ L 334, 1977)

---

car in European transport policies, an item which is not reflected in the CTP but is of central importance to transport policy and, even more important, to the lives of 320 million Europeans.

It is in the area of road transport, particularly international road haulage, that the CTP has been most active both in pursuing general objectives and in the implementation of detailed proposals and directives. Table 5.1 summarizes the most important initiatives which have been taken.

Twitchett (1981) claims that the application of the CTP to the commercial road vehicle sector illustrates many of the characteristics of the harmonization process (see Glossary) of the EEC itself. Article 100 of the Treaty of Rome on harmonization has exerted considerable influence via technical harmonization measures on the road transport sector. Twitchett discusses the harmonization of technical requirements under two different headings: axle-weight limits, and road-worthiness tests.

Axle-weight limits

This has been a major source of friction within the EEC pre-dating the entry of the UK, Denmark, and Ireland in 1973. The lack of agreement on lorry weights throughout Europe has been viewed consistently as a major obstacle to the establishment of a common market in the commercial vehicle sector. Without common specifications for axle weights, the EEC argued, it is difficult for lorry manufacturers in one community country to sell their products in other community countries. These difficulties are compounded by the organizational problems which face hauliers carrying loads around Europe either at less than capacity or in contravention of national regulations. The Council of Ministers accepted a maximum of 11 tonnes limit per axle and a total weight of up to 40 tonnes in 1972, but these levels were rejected by the UK, Denmark, and Ireland. Table 5.2 shows the current situation with vehicle weights and dimensions. The differences are indicative of the strength of different lobbying groups in domestic transport politics but also of long-standing differences in attitudes to road and rail movement of freight. In the early 1970s the Community was clearly divided into two camps: France, Belgium, and Luxembourg welcomed a 13-ton limit, and Germany, Italy, and the Netherlands a 10-ton limit. The 1973 entrants were in the latter camp, so shifting the political balance of the debate in the direction of the lower limit.

In 1984 the EEC revisited the theme of weight limits (85/3/EEC, OJ L 2/14, 3.1.85), claiming that differences between standards then in force 'are such as to have an adverse effect on the conditions of competition and constitute an obstacle to traffic between member states'. These maximum weights and dimensions laid down maxima of 40 tonnes for road vehicles with five or six axles, and 44 tonnes for a three-axle motor vehicle with a two- or three-axle semi-trailer carrying a 40-foot ISO container as a combined transport operation. Maximum axle weight was set at 10 tonnes for single, non-driving axles and 11.5 tonnes for driving axles. Ireland and the UK were specifically excluded from these provisions, enabling both countries to persevere with lower weight limits (though Ireland has now adopted the EEC maxima, leaving the UK isolated in this respect). The justification for the derogation in the case of Ireland and the UK was the poor condition of the road network and the need for improvement before higher weights could apply. This reference to poor road conditions certainly has a physical basis of truth but is also a face-saving device, particularly for the UK, which has a vociferous anti-

Table 5.2 Maximum weights applicable in the countries of Europe

	UK	Albania	Belgium	Bulgaria	Czecho-slovakia	Finland	Hungary	France	Italy	Nether-lands	Norway	Romania	Yugo-slovia
Height	4.2m	4.0m	4.0m	4.0m	4.0m	4.0m	4.0m		4.0m	4.0m		4.0m	4.0m
Width	2.5m	2.6m	2.5m	2.5m	2.5m	2.5m	2.5m	2.5m	2.5m	2.5m	2.5m	2.5m	2.5m
Length													
rigid	11.0m	11.0m	11.0m	11.0m	12.0m	12.0m	11.0m	11.0m	12.0m	11.0m	12.4m	11.0m	12.0m
artic	15.5m	12.0m	15.0m	16.5m	15.0m	16.0m	14.0m	15.0m	15.5m	15.5m	16.0m	15.0m	15.0m
roadtrain	18.0m	18.0m	18.0m	20.0m	18.0m	22.0m	18.0m	18.0m	18.0m	18.0m	18.0m	18.0m	18.0m
Axle weights													
single	10.5t	(3)	13t	10t	(3)	10t	10t	13t	10t	10t	10t(2)	10t	10t
double	20t	(3)	20t	20t	(3)	16t	16t	21t	14.5t	16t	16t(2)	16t	16t
Maximum gross weight	38t	38t	40t	(1)	38t	(5)	38t	38t	44t	50t	(1)	38t	40t

	Spain	Austria	Denmark	Portugal	Switzer-land	East Germany	West Germany	Sweden	Turkey	Luxem-bourg	Poland	USSR	Greece
Height	4.0m	4.0m	3.6m	4.0m	4.0m	4.0m	4.0m		4.0m	4.0m	4.0m	3.8m	3.8m
Width	2.5m	2.5m	2.5m	2.5m	2.5m	2.5m	2.5m	2.5m	2.5m	2.5m	2.5m	2.5m	2.5m
Length													
rigid	12.0m	11.0m	10.0m	12.0m	12.0m	10.0m	12.0m	(3)	11.0m	10.0m	11.0m	12.0m	12.0m
artic	16.5m	16.0m	15.5m	15.5m	16.0m	15.0m	15.0m	(3)	14.0m	15.5m	15.0m	20.0m	15.0m
roadtrain	18.0m	18.0m	18.0m	18.0m	18.0m	22.0m	18.0m	(3)	22.0m	18.0m	22.0m	24.0m	18.0m
Axle weights													
single	13t	10t	10t	10t	10t	10t	10t	10t	(4)	13t	10t	10t	13t
double	21t	16t	16t	16t	18t	16t	16t	16t	(4)	20t	16t		(6)
Maximum gross weight	38t	38t	44t	38t	28t	(1)	38t	48t	(4)	40t	38t	40t	38t

Notes: Weights given are the absolute maximum weights:
(1) total of axles.
(2) in Norway very few roads take these axles weights; lower limits apply.
(3) no restrictions.
(4) axle and gross weights are based on a complex formula.
(5) dependent on axle spacing.
(6) dependent on type of axle and spacing.

lorry lobby and an impressive array of evidence concerning the damage done by heavy goods vehicles. In a situation where rail has such a small market share of freight and the road system is in such a bad state of repair (the two factors being not unrelated), heavier lorries cause far more problems for politicians than in other European countries with a tolerable level of rail freight movement.

Road-worthiness tests

In 1976 the Council of Ministers issued a Directive on the 'approximation' of laws relating to road-worthiness for motor vehicles and their trailers (applies to all commercial vehicles over 3.5 tonnes laden weight) and all passenger vehicles with more than eight seats. Annual certificates of road-worthiness were made compulsory, with each country's national certification to be recognized by all other countries.

The harmonization of social measures is a separate theme in the evolution of the CTP and has also been the subject of a detailed investigation by a Select Committee of the UK Parliament (House of Lords 1983b), which concluded that the main regulations are 'unsatisfactory and too complex'.

The main EEC regulations under the heading of social measures are: (1) Regulation (EEC) 543/69 which limits the working hours of drivers of certain goods and passenger vehicles and fixes minimum rest periods; it also provides for keeping records to help enforce these limits; and (2) Regulation (EEC) 1463/70 which requires the installation of automatic recording equipment (tachographs) in most of the vehicles concerned.

Regulations on these matters do not represent a significant departure from national practices, where it has long been accepted that controls on the haulage and the road passenger industry are desirable. The regulations fall within that broad area of concern which covers health and safety at all places of work, and is essentially intended to protect workers from unscrupulous employers and to contribute to road safety by ensuring, as far as possible, that drivers are rested, refreshed, and not suffering from fatigue and up to the task of driving large vehicles over many hundreds of kilometres.

The EEC brings another dimension to this area of regulation. The drivers' hours rules (543/69) were introduced in 1969 to eliminate disparities likely to cause distortion of competition. Moreover, it is a very practical point of relevance to all member states that regulations do not differ enormously on a journey which traverses several different countries. The EEC harmonization in this area was seen as improving the market conditions for road haulage, improving social progress and improving road safety and, perhaps, as a way of outflanking those countries who might be tempted to use such regulations as a way of showing preference to railways (e.g. Germany).

Regulation 543/69 prescribes maximum continuous driving (4 hours), daily driving (8 hours), weekly (48 hours) and fortnightly (92 hours) driving and minimum daily rest (11 hours) and weekly rest (29 hours). In the UK this Regulation did not come into force in full until 1 January 1981. The rules are perceived as complex and, to a degree, unenforceable – a point that was made repeatedly in evidence to the House of Lords Select Committee on the European Community (House of Lords 1983b). The Regulations apply to road haulage and to passenger vehicles with more than fifteen seats (more than nine on international journeys). Clearly such regulations need a method of accurate record keeping if they are to be enforced and monitored by employees and employers alike. Originally Regulation 543/69 required a personal log-book for use by the crews. From 1 January 1978, this should have been replaced by the tachograph – and probably this one piece of technical harmonization in the transport field has had more effect on public perceptions of the EEC than any other measure through the years of argument between the EEC and the UK government over compliance with the regulation.

The EEC Regulation 1463/70 required that, from 1 January 1976, mechanical recording equipment should be installed and used in commercial vehicles throughout the Community, and substituted for the log-book. Tachographs were first used in Germany in 1953 for long-distance haulage – ironically, in view of the UK argument, at the request of the trade unions. Twitchett claims that the Dutch, French, and Belgians were suspicious of EEC rules in this field, but the tachograph gained acceptance quite rapidly as a means of protecting drivers from unscrupulous employers, whose scheduling and payment of drivers involved the assumption that long hours would be worked with few breaks: 'Tachographs are now apparently accepted on the continent as useful safeguards for both employers and drivers and there seems little regret over the passing of the individual control or log book' (Twitchett 1981:72). Britain did not accept Regulation 1463/70 and, in 1972, the Commission commenced legal proceedings against the UK government which were successful. The EEC Regulation 1463/70 applies in full in the UK from 1 January 1982.

The differences between the UK and other EEC member states on this issue are, in some ways, more interesting than the issue itself. Tacho-graphs are only a small part of technical harmonization, though they have achieved a significance beyond their humble origins. In Britain both governments and trade unions had a variety of arguments against them: their cost, the 'spy in the cab' argument, and unwelcome interference from the EEC in what is a matter for domestic determination. These views reflect a lack of experience in markets and distances which mainland Europeans have to contend with on a daily basis and also a degree of 'insularity' which still sits uneasily on Britain's membership of the EEC.

In 1985 the Commission of the European Communities made a proposal to Council to amend Regulations 543/69 and 1463/70 with a view to closer supervision and enforcement (there have been a large number of infringements), while emphasizing three basic requirements of legislation in this area: to produce healthier competition conditions; to further social progress (better living and working conditions for drivers); and to increase road safety.

However, it is clear from the text of the proposal that the existing regulations are regarded as too restrictive and some changes would produce a better deployment of vehicles, more streamlined operation of road transport (particularly long hauls), and increased productivity and profitability. This theme emerges in the detailed changes to driving hours, rest breaks, and so on, generally extending the hours and pro-viding for more flexible use of time (EEC Working Document A 2-9/85).

The social regulations for road transport specifically exclude any considerations of pay and remuneration, though Regulation 543/69 prohibits supplementary or bonus payment to wage-earning new members, if these are related to distance travelled and/or the volume of goods, unless they are of a kind which will not endanger road safety. Therefore, it is not possible to attribute a great deal of social progress in these regulations, as claimed by the EEC, since they ignore the relationship between pay and conditions of work. Low pay in the UK, and its necessary supplementation by shift-work and overtime, figure in trade union intransigence in the UK on the tachograph issue. Where conditions were better (especially pay), the trade unions took a different view. The conditions of work for transport drivers are still much worse than for many other groups of workers in terms of pay, unsocial hours, responsibility and accident risks, absence from home, conditions in the cab, etc. Social regulations have made little impact on areas of central importance to working conditions.

Quotas for carriage of goods by road

Though social regulation has figured prominently upon the public face of the CTP, just as much if not more debate and effort has surrounded the determination of quotas for the international carriage of goods by road.

The regulation and control of the road haulage industry at the domestic level has a long history in Europe. Button (1984) makes the point that at the time of formation the EEC was dominated by countries operating firm controls on the size of the industry, and that this naturally coloured their attitude to the activities of nationals of other EEC countries operating lorries either in the domestic haulage sector or in transit in third countries. Controls in the domestic sector arise from several concerns:

1.   a desire to protect railways
2.   a desire to limit environmental problems which are associated with road haulage (and limit damage to roads and hence road expenditures)
3.   a desire to ensure optimal (however defined) numbers of operators in the industry.

Controls on international quotas will reflect similar concerns with the additional complicating factor of equity tempered by political horse-trading. Put bluntly, it is obvious that an enlarged Community, even if it is partially successful at stimulating economic growth and development, will enlarge the market for international freight movement. Individual countries will want to maximize the benefits of this improved market opportunity to their own hauliers and to ensure that national interests are very strongly represented at the level of EEC decision making. This tendency could be expected to exacerbate existing national differences of opinion on control of road haulage capacity. Button (1984) provides a very useful summary of national attitudes to this sector of the transport industry, summarized in table 5.3.

The situation is a little more complicated than table 5.3 indicates because of the differences between haulage for hire and reward, and own-account haulage. This greater level of detail is covered by Button (1984) and will not be discussed here.

There are four ways in which road haulage between member states of the EEC is authorized:

1. Under bilateral agreements between member states which allow either free movement of goods or impose quotas on the number of movements
2. Under the Community quota arrangements which allow a limited number of vehicles to operate freely within the Community
3. Under the terms of the First Directive (1962) which prohibits any quota restriction on certain types of transport, and Directive 75/130/EEC which allows free movement of vehicles travelling partly by rail
4. Under European Conference of Ministers of Transport (ECMT) quota arrangements which are similar to the Community quota but apply to most ECMT countries (see Glossary). The ECMT permits

Table 5.3 National attitudes to road haulage regulations

France: Subject to capacity restrictions since 1934 when introduced as a railway protection measure; a zonal quota system severely limited long-haul markets particularly; in 1977 there was a relaxation of the controls partly in response to changed condition of the railways

West Germany: Strict controls on both long- and medium-distance zones; the 1952 Transport Act fixed the number of long- and medium-distance licences at 11,856 and 4,000 respectively; this number has been increased in line with demand but not at the rate of growth in GNP

Belgium: Operated a system of rigid long-distance haulage quotas up until 1960; this was abandoned, in 1960, and no limit was introduced to control the number of licences

Holland: The licensing regime introduced in 1954 applied to unscheduled haulage operations and incorporated a test of the hauliers' utilization of vehicles and profitability; this was tightened up, in 1974, and capacity could only be extended if new demand could be demonstrated; in 1975 total capacity was frozen and this policy has been reinforced by other measures in subsequent years

Italy: Haulage is subject to capacity restrictions, but not a rigid quota

Luxembourg: There is control of entry into professional road haulage by licence

UK: Abandoned quantity licensing in 1968, which dated from 1933 Act

---

Source: Button (1984).

are normally used for transits through the EEC to non-EEC countries rather than for movements within the EEC.

These are discussed in more detail below.

BILATERAL QUOTAS

There are five types (for greater detail, see Button 1984) and all are arranged directly between the two countries concerned. If a journey involves transit through a third country, then a permit will be needed from that country. This can produce problems, as Button notes: 'In 1970, for example, the UK had adequate permits for traffic to Spain but met problems in obtaining sufficient French permits for the necessary through movement' (p. 69).

Bilateral arrangements relate to numbers of journeys but there are some multiple-journey permits and some period permits, which do ease the administrative burden of the system if only a little. The UK has made bilateral road transport agreements with all the Community member states (1983). Six of them provide for no restrictions on entry (Belgium, Netherlands, Luxembourg, Denmark, Greece, and Ireland). In the other countries movement of goods vehicles must be authorized by permit. These bilateral arrangements within the EEC are recognized by the Commission and are subject to Directive 65/269/EEC, which prescribes a standard form of permit and other conditions.

COMMUNITY QUOTA

European Community multilateral permits (the quota system) are issued under Council Regulation 3164/76. They allow unlimited international journeys within the Community over one year; their significance is not great at the present time. The permits account for 5 per cent of the total intra-Community road haulage and much less than 1 per cent of total road haulage (which includes the domestic market) (House of Lords 1983b, para. 129).

Table 5.4 shows the Community quota system in its evolution from 1977-83. The quota system carries the hopes of the Commission of the EEC for liberalizing haulage movements. In evidence to the UK Parliament (House of Lords 1983b) the director-general of the EEC's transport division expressed the hope that the percentage of movements which takes place under the Community quota system could increase rapidly from 8 to 20 per cent, leading ultimately to the abolition of the system itself. The EEC is currently working towards 'a market without quantitative restrictions by 1992' (EEC 1985b, para. 2.3.1).

The number of quota authorizations has increased from 4,038 in 1984 to 5,268 in 1985, and to 7,437 in 1986, stemming from a Council decision of December 1984 to increase the number of authorizations by 30 per cent in 1985, and by 15 per cent for each of the following four years. An additional boost occurred in 1986 due to the accession of Spain and Portugal to the Community.

Each state has a quota, as can be seen in table 5.4, and is responsible for its own allocation procedures. These permits allow transport between member states but not cabotage (see Glossary). Certain road traffic within the EEC (amounting to about 35 per cent of the total) are fully liberalized and thus exempt from the need to hold a permit. The list of exemptions covers frontier traffic, luggage, removals, funerals, works of art, fairs and exhibitions, and carcases for disposal, together with a range of other highly specific commodities which are not very significant in overall road haulage terms. The importance of the Community quota system, in terms of actual traffic carried, can be seen in table 5.5.

The size of the allocations to each member state has frequently been a source of controversy, with strong pressure coming from peripheral states for an increase in their quotas. The UK regards itself as particularly disadvantaged because of the short sea crossing and the importance of lorry haulage via ferries to continental destinations. The rapid growth of road haulage and the relative decline of international rail transport has sustained a demand for more quota licences, as well as worries about the financial position of railways.

The evidence provided to the House of Lords Select Commitee (House of Lords 1983b) by the UK Road Haulage Association illustrates some of these points:

> the British share of this quota has not increased to the extent that the RHA would wish ... In 1970, two million tonnes of goods moved by road via 'Roll on/Roll off' services to and from the UK. In 1981 the total was 17 million tonnes.

Later in their evidence the RHA say:

> A progressive increase in the EEC quota is desirable because the multilateral nature of these permits encourages the more productive use of vehicles and the development of genuine, integrated

Table 5.4 Community quota, 1977–83

	1977	1978	1979	1980	1981	1982	1983
(a) Number of permits							
Belgium	265	318	348	413	413	434	434
Denmark	169	203	229	286	286	305	305
France	409	491	533	627	627	656	656
Germany	427	512	567	689	689	727	727
Greece	–	–	–	–	76	88	88
Ireland	50	60	65	76	76	88	88
Italy	319	383	432	539	539	567	567
Luxembourg	70	84	91	106	106	111	111
Netherlands	382	458	502	597	597	626	626
UK	272	326	355	418	418	436	436
Total	2,363	2,835	3,122	3,751	3,827	4,038	4,038
(b) Share of permits (%)							
Belgium	11.2	11.2	11.2	11.0	10.8	10.7	10.7
Denmark	7.1	7.1	7.3	7.6	7.5	7.6	7.6
France	17.3	17.3	17.1	16.7	16.4	16.2	16.2
Germany	18.1	18.1	18.2	18.4	18.0	18.0	18.0
Greece	–	–	–	–	1.9	2.2	2.2
Ireland	2.1	2.1	2.1	2.0	1.9	2.2	2.2
Italy	13.5	13.5	13.8	14.4	14.1	14.0	14.0
Luxembourg	3.0	3.0	2.9	2.8	2.8	2.7	2.7
Netherlands	16.2	16.2	16.1	15.9	15.6	15.5	15.5
UK	11.5	11.5	11.4	11.2	11.0	10.8	10.8
Total	100	100	100	100	100	100	100

Source: House of Lords (1983b).

Table 5.5 Percentage of total traffic carried under Community authorizations, 1981

	B&L	D	F	WG	I	N	RI	UK
Belgium and Luxembourg	–	5	2	16	22	1	2	2
Denmark	5	–	7	12	14	2	9	8
France	2	5	–	30	4	7	15	4
West Germany	11	8	26	–	73	12	1	1
Italy	26	15	4	80	–	16	*	*
Netherlands	–	2	–	11	18	–	13	3
Republic of Ireland	2	18	19	2	*	7	–	*
UK	2	11	5	1	*	23	*	–

*Not calculated due to inadequate data.
Source: Button (1984).

Community haulage. It is not felt that the Commission's current proposals on the distribution of these permits between States takes adequate account of the difficulties faced by hauliers with sea crossings to the other states in the community. The proposed formula would unfairly limit the UK's share of increases in the quota. (House of Lords 1983a:107)

The Commission's proposal for allocating the 1982 quota, which worried the RHA, was based upon each member state's existing share of trade and the use made of licences, taking into account the distance of some members from the centre of the Community (defined as the Frankfurt region). This proposal was rejected, but Ireland and Greece received more permits to compensate for their geographical position. Suggestions for the 1983 quota included the use of forecasts of road haulage activity as a basis for allocations. A 4.3 per cent forecast increase in tonnage carried by road haulage was equated with a quota increase of 174 licences. West Germany and Italy blocked this increase, so that the 1982 level was retained for a further year.

In 1983 the EEC Commission came up with a new set of proposals which would, if adopted, remove the need for laborious annual haggling over the size of the quota, as well as remove several economic inefficiencies and move the system of controls on international road haulage further towards the 'liberalized' goal (COM (83) 58 Final, EEC, 1983).

The Commission regards the present system of national and bilateral authorizations and regulations on capacity as 'costly, cumbersome and economically inefficient' and clearly does not like any system of controls (even the Community quota itself) because these: 'constitute(s) a considerable departure from the organisation of transport in accordance with market economic principles' (House of Lords 1983a: 51).

The costs and the economic inefficiencies referred to in the Commission's document arise from the restriction on the amount of professional road transport, when the customers might use more, if it were available and from the desire of some member states to protect their railways. A further inefficiency arises in the relationship between professional road haulage (for 'hire and reward') and own-account transport:

> Our investigations amongst the manufacturing industries in the Community indicate to us that where they use road transport, professional road haulage, to start with, if they are deprived of that by some restriction or other they tend to move not to the railways but to own-account transport. (House of Lords 1983a: 93)

Because 40 per cent of own-account journeys are empty, it is argued not only that this tendency does not help railways, but that it actually increases environmental damage and energy consumption. Thus liberalized transport can be represented as environmentally sound.

The Commission, in its 1983 document, identifies three sources of opposition to its attempts to improve market efficiency:

1. Governments who want to protect railways
2. Governments whose countries serve as transit routes for traffic from which they benefited little or not at all
3. Road hauliers already in possession of authorizations who want to avoid more competition.

The realization that opposition does exist, does not deter the Commission from proceeding and indeed the 1983 proposals could be interpreted as having benefited from the development of a sharper political instinct. The proposals are more far-reaching than previous efforts in this area and they link together road and rail considerations.

The Commission has set out its plans to increase the proportion of traffic moving under Community authorization by linking the growth in the number of Community authorizations with overall road traffic growth, but goes further: 'Parliament has advocated a longer-term solution to this issue and it is possible, for instance, to envisage a five-year Community agreement which would result in an increase of

Community licences each year by X times the rate of increase of total community road traffic in the previous year' (EEC 1983b, COMM (83) 58 Final). Other proposals include:

1. The Commission will identify and propose to liberalize fully, or at least partially, certain specific types of traffic which for commercial or economic reasons may best be suited to road transport
2. A system of compensation for transit states via infrastructure charges and revenues; worries on the part of transit states have been an obstacle to the development of the CTP and these should be overcome without adding to frontier delays or changing the 'nationality principle' of taxation
3. Measures to improve the competitive position of the railways 'consisting in particular of measures designed to solve the infrastructure problem and to remove any obstacles that exist to a closer co-operation between them'
4. Removal of unnecessary bureaucratic delays at frontiers and speeding up frontier crossings; in particular, control on fuel in tanks should be abandoned
5. Harmonization of weight and dimension of community road vehicles
6. Introduction of a system of reference tariffs for road haulage
7. Changes to the definition of own-account transport to include services provided under long-term contract exclusively to a particular customer
8. Suggestions for a first directive on the adjustment of national taxation systems for commercial vehicles; this is intended to harmonize the conditions of competition within and between modes of transport
9. Suggestions for the review of Regulations 543/69 and 1463/70 on social regulations for road transport.

The EEC's approach to the whole question of Community quotas is riddled with internal inconsistencies. The CTP is avowedly free market in its objectives and committed to liberalization, yet finds itself in the uncomfortable position of supervising and shepherding an enormously cumbersome bureaucratic system, which has all the characteristics of national policies of which the EEC is so critical and which are deemed to be damaging to efficiency. Clearly it would be difficult for the EEC to sweep away fifty years of regulations at the national level which has put road haulage into this peculiar position, but more difficult to explain is the lack of progress on liberalization in the transport sector, or the mental gymnastics which put the case for increased quotas and liberalization as environmentally sound and of little consequence to railways. This is poor judgement and the CTP has failed to grasp the basic relationships between the ways in which road and rail infrastructure is funded and organized and the degree to which each is used. Vague assertions that rail's competitive position will be improved are of little value when set against the rapid growth of road haulage and the strenuous efforts which are made to ease the passage of the lorry across national frontiers.

The EEC does not face the same kind of lobbying pressures as do national governments, and progress towards liberalization and harmonization is not fast enough for many organizations representing lorry operators in the UK. The Freight Transpsort Association in the UK has set out its criticism of the CTP (FTA 1986), together with its demands for heavier lorries, removal of weekend driving bans in some European countries, liberalization of cabotage, increases in size of quotas and stronger action against member states who channel funds into their railways outside the scope of EEC regulations which are framed to control rail subsidies.

The FTA argue that industry requires the provision of adequate infrastructure, particularly road infrastructure and a major road network free of tolls and toll-booth obstructions. They also argue that the free movement of goods is impeded by:

1. disparities between VAT and excise duties
2. collection of statistics, CAP levies, excise duties and VAT at frontiers
3. failure to adopt a Community-wide system of postponed accounting as proposed by the fourteenth VAT Directive
4. requirement for a guarantee to move goods under the Community transport system.

The FTA are clearly at one end of a spectrum of alternatives, which sees the CTP as a major force for pushing home the competitive dominance of the lorry in European freight movement. The CTP policies are, in any case, already skewed in this direction, and the FTA's endorsement of many aspects of recent CTP initiatives is a clear indication of this bias.

6 The car

The car is, without doubt, one of the most persuasive and intrusive technologies of the twentieth century, and one which is still little understood in terms of its effects on quality of life, economic welfare, spatial structures in society, energy, and the future of the city. Many of these issues have been discussed by McShane (1984), but there is little consensus on the issues raised by use of the car. Table 6.1 is an extended checklist of car-related issues based on the work of the 'Future of the Automobile' project, described by McShane.

The range and complexity of many of these topics is perhaps sufficient explanation for the lack of success on the part of national transport policies and all but a few of the relatively minor issues are beyond the competence of a Common Transport Policy (Whitelegg 1979). Nevertheless, the EEC has pursued policies, described earlier, which will increase motorized traffic and has failed to recognize the fundamental incompatibility of pursuing greater mobility as a policy objective, while at the same time legislating for exhaust emissions and organizing a 'road safety year' (1986). Lack of attention to social objectives within the CTP, particularly the severe disparities between income groups and subgroups in their mobility, is a serious omission from the CTP agenda, when the EEC in general claims a social purpose (e.g. the Social Fund) and claims a wide-ranging interest in the condition of life for Europeans (e.g. the European Foundation for the Improvement of Living and Working Conditions). This latter interest has spawned a study of commuting and its implications (Pickup and Town 1983), which is a good example of the possibilities for achieving Europe-level understanding of complex transport phenomena as an essential precondition to some form of action. The absence of even a rudimentary EEC evaluation of the societal and environmental context within which the automobile operates does have the effect of producing isolated policies which fail to achieve their full potential or may even, in extreme cases, conflict with

Table 6.1 Car issue checklist

1. Speed limits
2. Drink-driving laws
3. Seat-belt laws
4. Vehicle inspection and vehicle standards
5. Driving licensing, training, and monitoring
6. Parking provision, standards, and costs
7. Fuel prices, energy policies, and car use
8. Government taxation of car ownership and car use
9. Consumer protection and information
10. Infrastructure (track) provision for cars
11. Balance of fiscal support for private and public transport
12. Noise standards
13. Vehicle emission standard
14. Design of residential areas for protection of young, elderly, and other 'at-risk' populations
15. Impact of car on urban design and costs of servicing urban areas
16. Distribution of car mobility through different population groups, e.g. women, the elderly, ethnic minorities
17. Effects of cars on walking and cycling – through perceived dangers of unprotected road users
18. Place of cars in contemporary industrialized societies, dependence on car manufacturing, and role and influence of car manufacturers on national economy and jobs and transport policy
19. Role of car in changing the function of city commuting, cross-suburban movements, de-urbanization, re-urbanization, sub-urbanization
20. Social effects of the car in polarizing society: mobility rich vs mobility deprived

one another. Thus technical standards for vehicles may conflict with road safety objectives if they ignore the relationship between design and speed capability. Attempts to draw attention to road safety and accident reduction can only make sporadic progress against the volume production of high-performance cars and little decisive action to limit their speed. Similar conflicts with energy and environmental quality objectives will result if policies towards the car, and its control and use, are not developed within a clear enough framework of goals and objectives. At the moment, there is no such clear framework.

Car ownership in Europe varies enormously, as is shown in table 6.2. Clearly West Germany is a highly motorized society, while Greece, Spain, and Portugal are at the other end of the scale. Such different

Table 6.2 Car ownership in the EEC, 1984

Country	Motor vehicles* (000s)	Motor vehicles per 1,000 inhabitants
West Germany	25,378	416
Belgium	3,300	335
Denmark	1,440	283
France	20,800	380
Greece	1,151	118
Britain†	16,104	293
Ireland	711	203
Italy‡	20,450	359
Luxembourg	146	· 398
Netherlands	4,772	330
Portugal§	1,346	144
Spain	8,874	230

*Excludes buses, lorries, mopeds, and motorbikes.
†Excludes Northern Ireland.
‡1983 data.
§1981 data.
Source: *Statistisches Jahrbuch 1986 für Bundesrepublik Deutschland*, Wiesbaden, Statistisches
Bundesamt, 1986.

experiences inevitably produce different attitudes to control and
management of the car, to the provision of roads and other facilities to
sustain growth in car ownership. The EEC is indeed uneven in the rate of
development of motorized societies and is unlikely to see much agreement on such basic transport issues as car ownership and use. Table 6.3
gives some idea of these disparities as revealed in differences in levels of
road provision. In terms of motorway provision, Germany and the
Netherlands are at the top of the league, with the UK and Spain far
down. Ireland has no motorways, while Portugal has only 0.013 km of
motorways per 1,000 people. Main roads provide a somewhat different
picture, with small countries with small populations coming out with
greater levels of provision. Thus, Luxembourg and Belgium are ahead of
the rest in these terms. Once again, the UK comes low in the table, and
particularly when compared with Denmark, Greece, and Ireland with
almost four times the level of provision. In terms of road space per capita,
the UK ranks behind those countries with which comparisons would
normally be made (Germany, France, Italy), and behind some smaller

Table 6.3 Road provision in EEC countries, 1982 (length in kilometres)

Country	Length of motorway	Length of main roads*	Motorway (km) per 1,000 people	Main road (km) per 1,000 people
West Germany	7,919	32,239	0.128	0.524
France	5,290	28,130	0.097	0.516
Italy	5,901	45,147	0.103	0.794
Netherlands	1,841	2,760	0.128	0.192
Belgium	1,388	11,909	0.140	1.208
Luxembourg	44	926	0.120	2.530
UK	2,765	12,871	0.049	0.228
Ireland	–	2,629	–	0.749
Denmark	516	4,129	0.100	0.807
Greece	91	8,689	0.009	0.877
Spain	2,072	–	0.054	–
Portugal	132	–	0.013	–

*Most important routes which are not motorways: Germany, Bundestrassen; France, routes nationales; Italy, strade statoli; Netherlands, belangrijke rijkswegen; Belgium, routes d'état/rijkswegen; Luxembourg, routes d'état; UK, trunk roads; Ireland, national primary roads; Denmark, hovedlandeveje; Greece, national roads

countries whose dense network of roads serves a transit as well as a local function. Unfortunately for transport planning and policy making, car ownership and use cannot be discussed in isolation from other societal factors. The demand for car travel is itself a function of both long- and medium-term changes in the physical land use environment and, in particular, of the provision of road space (table 6.3). The traffic generation effects of new road construction have been studied in London (Beardwood and Elliot 1985) with results which show clearly that new road construction, the provision of more road space, generates new traffic which deprives existing roads of relief and fuels the next round of increase in car use and demand for yet more road space. The fiscal environment is also important. Car tax, fuel tax, the perceived cost of car use, together with subsidies to private car ownership and use, will naturally fuel the demand for road space and car use. Whitelegg (1984b) has discussed this with respect to the UK company car sector, and Kutter has discussed the Danish experience of car taxation:

> [Taxes and duties] in Denmark today are thus four times higher, comparing the medium size car in Denmark with the car in other

European countries. From 1971 to 1976 expenditure on cars rose sharply by 60% while the fares of public transport become cheaper. Due to this development private car traffic decreased by about 10% while public transport increased sharply by about 50%.
(Kutter 1987: 56)

The car market itself is heavily influenced by national policies towards the support of the motor car industry which has traditionally assumed a regional policy and central economic policy role in job creation and attraction of inward investment (as the excitement over the Nissan plant and its eventual location in north-eastern England). Some idea of the importance of the car industry can be seen in the study, 'The Future of the Automobile' (Jeschke and Kunert 1984). When discussing Germany, the authors conclude that 1.8 million people directly or indirectly depend on the demand for cars for their work. This is 7 per cent of the labourforce. In 1980 the car industry produced a value of 106 billion Deutschmarks, representing 7.3 per cent of total industrial production; 3.9 million vehicles were produced, of which 3.2 million were private cars and 54 per cent exported, making the car the most important export article of Germany. Table 6.4 shows car production in 1984 for a number of European states.

Motor vehicle manufacturing is of obvious importance to the economies of those states involved in this industry – and the importance will tend to grow over time as manufacturing, as a sector, shows signs of decline. Few industrial sectors have the massive governmental support and fiscal encouragement which the car enjoys, this guarantees a market for the product which in spite of recession and fuel price fluctuations is remarkably stable. Motor cars in western societies share this exalted status with defence industries, and in both cases the existence of a product and a self-perpetuating technological dependence is supported by an ideology

Table 6.4 Motor vehicle production in selected European countries, 1984 (thousands)

West Germany	3,783
France	2,713
Britain	1,029
Italy	1,439
Spain	1,174

Source: ECMT (1986a).

which justifies the need and excludes alternatives for satisfying that same need.

The automobile industry has a significant influence on important aspects of the car and its impact on society; it has involved itself in the debate on speed limits in Germany (against autobahn restrictions), and in other countries the industry exerts itself in those directions which will favour its products and their sale. In advertising the industry extols the virtues of speed and design features which are not in the interests of safety (Volvo is an exception). In basic design, particularly engine size, power, and speed, the automobile industry has adopted an irresponsible attitude in its development and production of vehicles which can achieve higher speeds - often beyond legal maxima. In 1979 the proportion of vehicles capable of at least 140 m.p.h. was about 50 per cent compared with approximately 10 per cent in 1969.

Motorization is a very clear result of very clear political processes. Yago (1984), in his thorough study of the USA and Germany, shows this in some detail. The activities of government, transit companies, car manufacturers and related industries, and political ideologies combine to represent a powerful force in society in one direction - as far as the achievement of mobility and satisfaction of transport needs is concerned. Discussing the development of motorization in Germany, Yago identifies the National Socialist era (1933–45) as of particular importance:

> Motorisation, mobilisation and militarisation of the society and the economy merged and became central to Hitler's economic and social policies. Motorisation would stimulate the economy and create jobs. The society could be mobilised to create a highway transportation system that would unite the Reich. Motorisation was important for creating a modern military whose expansion would spur economic growth and instil social discipline. In short, motorisation was simultaneously a military, ideological and economic strategy in the Nazi drive to restructure German society.
> (Yago 1984: 37)

While the identification of such an urgent need to restructure society may be absent in other European countries, the basic point made in this reference to Germany still stands: motorization as an instrument of national, social, and economic policy is vigorously pursued and supported by powerful alliances.

The consequences of motorization are not primarily the concern of those who support this technology. In this chapter we will examine road traffic accidents and the related question of speed limits in some detail. In chapter 11 environmental – particularly pollution – aspects are reviewed.

Road traffic accidents

Road traffic accidents (RTAs) are a particularly well–documented consequence of motorization. While accidents occur in all modes of transport, including railways, no mode approaches the importance of the motor car in the scale of deaths and injuries caused to vehicle occupants, pedestrians, and other unprotected road users. The scale of the problem is such that if RTAs were regarded as a disease in the normal sense of the word, the public outcry and resources deployed to combat the problem would be such as to produce very rapid changes in the way road accidents are regarded and in their quantitative impact.

In the Europe of 'the Ten' (i.e. excluding Spain and Portugal), RTAs have killed more than half a million people since 1970 and injured over 15 million. In 1982, in 1,144,032 separate accidents, 1,514,900 people were injured and 45,700 were killed. These data are illustrated in figures 6.1 and 6.2.

Figure 6.1 shows a gratifying downward trend in RTAs which, while not great, is certainly in the right direction. Evidence from Britain advises caution on the interpretation of these trends. The decrease in Britain, from 5,934 in 1982 to 5,445 in 1983, conceals some interesting variations within the aggregate totals. Motor-cyclist casualty rates increased by 3 per cent after a ten-year decrease, pedal cyclists' casualty rates increased by 9 per cent, and child pedestrian deaths increased by 17 per cent. In the UK this point was made forcefully to a House of Commons Select Committee in 1985:

> Each of the non-motoring modes of travel has actually become more dangerous. Deaths and serious injuries to older children and young adult pedestrians (10–19 years) increased from 2208 in 1958 to 5059 in 1982. The corresponding rate per 100,000 individuals in this age band also doubled. (Vol. 3, p. 99)

Table 6.5 shows something of the variation between European countries in 1982 in the incidence of RTAs.

In this manner, with numbers injured or killed standardized by reference to the country's total population, it is possible to see that France has the highest mortality rate and UK the lowest. For males and females, France is also the most dangerous country and the UK the safest, a

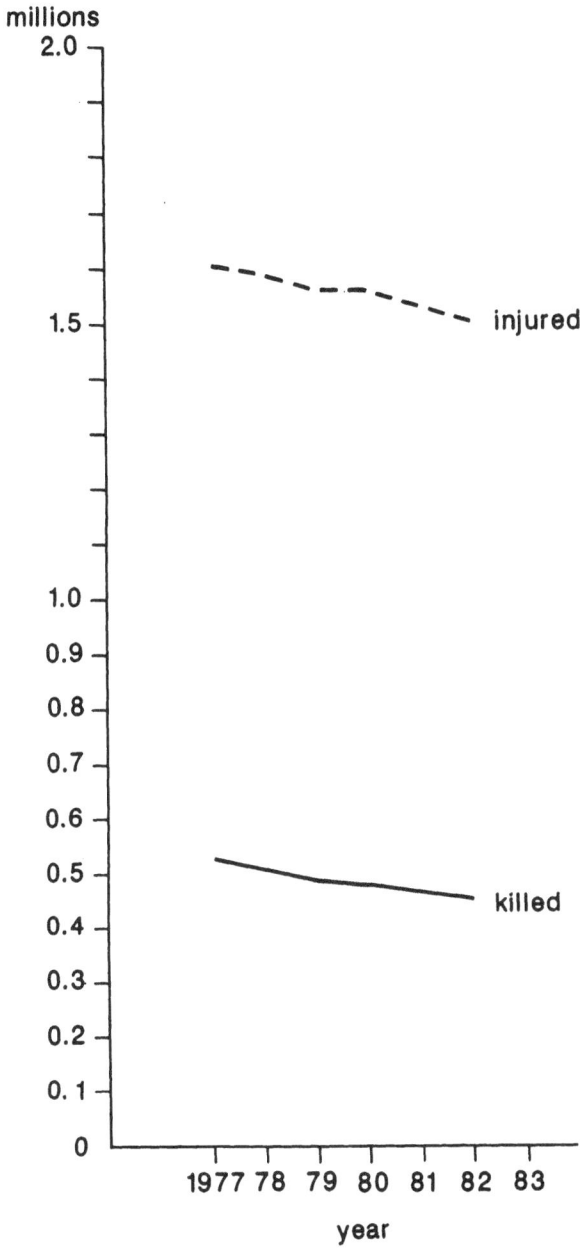

Figure 6.1 Road traffic accidents in the ten countries of the EEC

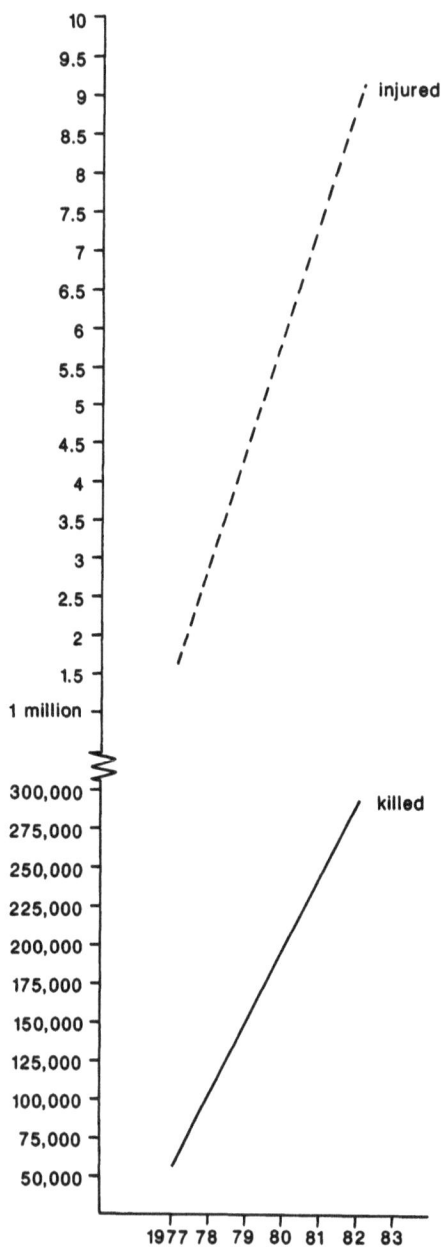

Figure 6.2 Road traffic accidents in the EEC (cumulative totals)

Table 6.5 Road traffic casualties per 10,000 inhabitants

Country	Total killed	Males killed	Females killed
Germany	1.9	2.8	1.0
France	2.2	3.3	1.2
Italy	1.3	2.1	0.6
Netherlands	1.2	1.7	0.7
Belgium	2.1	3.1	1.1
Luxembourg	2.1	3.1	1.1
UK	1.1	1.6	0.6
Ireland	1.5	2.3	0.7
Denmark	1.3	1.8	0.8
Greece	1.6	2.5	0.7

Source: Eurostat, Transport Communications, Tourism, 1970–83.

position shared with Italy in the case of female deaths. It is difficult to be accurate about the reasons for variations between countries in these rates, though it is somewhat easier to explain male–female rates. In the case of women, their traditional role in society which excludes them from many of the activities and characteristics of male roles, ensures that they have fewer driving licences, less access to a car even if they hold a licence, and certainly much reduced opportunities to embark on a number of different journey purposes associated with work, leisure and recreation, etc. (Oliver 1986). This reduced exposure to a hazardous environment, then, produces lowered casualty rates. In the case of country by country variations, the explanation lies in a complex mixture of legalistic, attitudinal, and environmental factors that would be difficult to disentangle. Variations in speed limits, public transport quality (which influences the relative importance of different modes, e.g. walk, car, bus), population densities, legislative initiatives on car maintenance, drunken drivers, seat belts and road behaviour, and the vigour with which improvements to traffic environments are pursued, all influence the outcome.

A more basic problem with international comparisons of road traffic accidents concerns the definitions that are used. In Spain, in 1983, 4,666 deaths were recorded in RTAs, but had the UK definition of an RTA fatality been used, i.e. death within 30 days, rather than 24 hours, then the Spanish figure would have been more like 6,000.

Inconsistencies in the recording of deaths is not the only problem.

Table 6.6 Scale of under-reporting of road traffic accidents

	Known to police	Not known	Σ	Percentage 'hidden'
(a) *Birmingham*				
Deaths	15	0	15	0
Serious injuries	297	64	361	18
Slight injuries	535	289	824	35
Total	847	353	1,200	29
(b) *Berkshire*				
Deaths	89	0	89	0
Serious injuries	1,205	327	1,532	21
Slight injuries	1,340	680	2,020	34
Total	2,634	1,007	3,641	28

Source: Analyse von Unfalldunkelziffern
    W. Siegener
    W. Lenhart
    IVT Ingenieurbüro für Verkehrstechnik Gmbh Karlsruhe.

Table 6.7 Traffic casualties, by means of transport, 1982

	Number	%
Total casualties*	1,560,598	100
Pedestrians	203,135	13.01
Drivers and passengers in cars	802,592	51.42
Goods vehicles	44,218†	2.83
Buses, coaches, and trams	21,737†	1.39
Motor-cycles	192,169	12.31
Mopeds	156,458‡	10.02
Cycles	133,677	8.56
Other vehicles	6,613	0.42
Air Rail	2,648§	

*Not including air and rail.
†Excludes Luxembourg.
‡Excludes Luxembourg and Ireland.
§Includes passengers, railway employees, and others.

Today advances in medical science mean that more accident victims are surviving beyond 30 days, resulting in an underestimation of deaths should these patients eventually die. There will also be variations in the degree to which accidents with injuries are brought to the attention of the police. In the case of injury accidents, some reconciliation of police accident data with hospital data may well be necessary to produce an accurate figure. Table 6.6 presents results from a UK study on the scale of under-reporting of RTAs.

Pedestrian casualties account for 13 per cent of all casualties in the ten EEC states (1982). Table 6.7 details traffic casualties by their means of transport. There are marked variations between member states in the incidence of pedestrian casualties. Table 6.8 summarizes this succinctly, though the data are now a little out of date. The differences in performance are quite marked. If Germany could 'trade up' to the Netherlands rate for all pedestrians in 1979, it would have killed 1,150 instead of 3,167, a reduction of over 2,000 deaths.

The effects of road traffic accidents are not confined to deaths, injuries, and emotional distress caused to relatives of the victims. There

Table 6.8 Pedestrian, and child pedestrian, fatal road casualty rates in OECD countries, 1979 (deaths per million population)

All pedestrians		Pedestrians aged 0–14 years	
Country	*Rate*	*Country*	*Rate*
Netherlands	18.7	Sweden	9.7
Sweden	21.4	Finland	13.3
Norway	23.8	Netherlands	18.3
Japan	24.9	Spain	21.1
Denmark	28.2	Denmark	24.6
Finland	31.9	Japan	25.4
Spain	34.2	France	25.9
USA	37.4	Norway	26.3
Canada	37.6	USA	28.1
UK	39.0	UK	32.6
France	39.3	Canada	34.3
Belgium	50.4	Belgium	35.5
Switzerland	51.0	Switzerland	38.6
West Germany	51.5	West Germany	40.7
Ireland	67.4	Ireland	48.3

Source: OECD (1983a).

are economic effects in terms of lost production and of the cost of maintaining accident and emergency services associated with the aftermath of accidents. Additionally, there are costs associated with the administration of the legal system and with the business of insurance claims related to road traffic accidents.

Most EEC countries have methods of estimating these costs, even though the procedures involved are not likely to be precise and will involve judgements which are difficult to verify empirically. In the UK, in 1983, road traffic accidents cost £2,380 million. Grossed to a European scale, these costs represent a serious drain on public expenditures in Europe and resources which could be deployed elsewhere if the level of RTAs could be significantly reduced.

Road traffic accidents have other serious implications which cannot be incorporated into a system of costs and benefits. Road traffic accidents represent a major negative aspect of urban living in European cities. Of the 1.144 million accidents in Europe (ten states) in 1982, 831,000 or 72.6 per cent were within built-up areas. In the UK, roads in built-up areas produce 1.13 times more fatalities, 1.91 times more serious injuries, and 3.08 times more slight injuries than roads in non-built-up areas (Department of Transport 1984). Evidence was presented to the UK House of Commons Transport Select Committee which showed that 95 per cent of pedestrian accidents occur in built-up areas, and 95 per cent of child pedestrian accidents under the age of 5 years, occur within a quarter of a mile of the child's own home (House of Commons 1985: Vol. 2, p. 39). Road traffic accidents on this scale produce fear and anxiety principally among parents of young children and elderly citizens, whose physical mobility characteristics produce particular problems in a heavily trafficked environment. Fear and anxiety are difficult to measure in any way which would enable comparisons to be made between European countries. Appleyard has put the problem of developed urban environments succinctly:

> The protection and creation of livable streets is not simply a matter of increasing the comfort or safety of urban living. The street has other functions. As the place where most children grow up, it is a crucial mediator beween the house and the outside world, where the child learns to confront strangers and environments on his own. It should be a receptive and reasonably safe environment that the child can explore, manipulate and use as a setting for all kinds of activities. (Appleyard 1981: 9)

The European urban environment, with few exceptions, does not measure up to these standards.

Policy towards reduction of RTAs

Road traffic accident reduction is an established part of all European countries' transport policies, though there are serious problems with the usual methods which are employed in the fight against accidents (White-legg 1982). Table 6.9 summarizes the main areas of RTA reduction policy.

Table 6.9 Approaches to reducing road traffic accidents

Educational
(a) Measures aimed at children, particularly through schools and parents, to instil ideas of 'good' road behaviour.
(b) Campaigns aimed at the general public, e.g. on drink, speed, seat belts, etc.

Engineering
Measures aimed at producing as safe a physical environment as resources will permit – road surface, cambers, lighting, sight lines, eradication of 'black spots', traffic management methods, improved signals, and the building of new roads

Legal
Establishment of norms which are enforced by penalties to improve behaviour, e.g. speeding, drink-driving, seat belts, 'carelessness'

Enforcement
Closely related to legal category but involving the deployment of police resources and improving the odds of detection and prosecution; also involves courts and their attitude to RTAs *vis-à-vis* other crimes against the person

Land Use planning
Involves design of an urban environment at different scales so as to produce safe journeys, reduce journey length, and give special attention to residential environments where children and the elderly live; also involves detailed urban design on the Dutch *woonerf* model

Public transport
Uses the high-quality safety provided by these modes to shift journeys from unsafe modes to safer modes

EDUCATIONAL MEASURES

Educational measures are widely practised throughout Europe, though it is difficult to evaluate their effectiveness. In England, 'the green cross code' has been standard training for young school children for many years, and the Greater London Council introduced its own variation of measures to involve parents and children on a continuing basis in road user education. In evidence to a House of Commons Select Committee (House of Commons 1985) some doubt was cast on the training of one particular class of road user, the motor-cyclist: '... the sample of those who had been through the training scheme had a higher accident rate per mile than those who had not and [that] the difference was statistically significant' (Vol. 1, para. 54). There is also some evidence that campaigns directed at driver behaviour, as in 'drink drive' campaigns, have an initial beneficial effect but one that cannot be sustained on a long-term basis.

ENGINEERING MEASURES

It is clear that the design of roads, bends, junctions, lighting, etc. will have an influence on accident levels if that design is sub-standard or unsuited to the conditions in the vicinity of the accident site. Accident specialists are accustomed to identifying 'black-spots' where the level of accidents exceeds a given threshold in a certain time period. In the city of Darmstadt in West Germany (population 137,000) some fifty black-spots have been identified where there are more than five accidents of a certain type in one year. These are then scheduled for various types of remedial treatment. There are many problems with this approach (Abbess, Jarrett, and Wright 1981; Boyle and Wright 1984). First, there is a real danger that the benefits of such an approach are exaggerated by the tendency for accident levels at sites to vary without treatment, and since extreme sites (i.e. more accidents than 'normal') are chosen, these are more likely to show an improvement anyway, regardless of treatment. This 'regresssion-to-mean' effect is difficult to quantify. Secondly, if a black-spot is eradi- cated, there is a suggestion that this may have the effect of relaxing the concentration and awareness of drivers, so that in the vicinity of the now- removed black-spot, other black-spots may occur. This is the accident migration effect, which is now the subject of some debate. As in the education of road users to improve their behaviour, there is some uncertainty that traditional methods are as effective as was once thought – or indeed are as efficacious as alternatives which might be available.

Engineering solutions in cities also include close attention to junction design and layout, signals at intersections, and so on, and there is ample evidence to justify close attention to new junction design and the application of sophisticated, computer-controlled systems of signals.

Engineering solutions may also involve the construction of new roads which, it is generally argued, are safer than existing roads. The construction of new roads and motorways is included in the 1973 Road Safety Programme of the Federal Republic of Germany 'as an essential prerequisite to promoting road safety'. The traffic generation effects of new roads casts considerable doubt on this argument (Transport 2000, 1986) and it is by no means clear whether new roads provide a net gain or loss to road users (particularly unprotected road users) and their safety.

LEGAL MEASURES

Road transport is heavily circumscribed by legislation and regulation both in the details of how road space is to be used (e.g. speeds, one-way streets, stop and give-way instructions) and in the condition of the vehicle (e.g. the British MOT and the German TUV tests). Added on to this system for the organization of the track and the vehicle are prohibitions on the driver (e.g. in the consumption of alcohol and the wearing of seat belts). In all these areas road traffic laws are complex in all member states and are unlikely to be candidates for harmonization within the framework of the Common Transport Policy. Driver licensing is one exception, but this is not significant compared with the totality of legislation and regulation which affects the private motorist. A further difficulty lies in the philosophical conception of punishment and deterrence. Road traffic law infringements, when detected, may be regarded lightly by the courts, may attract fines, cautions, or imprisonment, or various forms of attendance at centres for such offenders. There is little consensus of opinion on what form of 'punishment' suits this kind of crime, and indeed how different forms of punishment may vary in their role as deterrents or behaviour modifiers.

ENFORCEMENT MEASURES

This category of potential actions which have a bearing on reducing RTAs is part of the legal process but it has distinctive elements which merit separate consideration. For historical reasons and for operational convenience, road traffic law is enforced by the police in European countries. Sometimes there are separate police forces, but usually different departments of one police force will be attached to traffic duties. This may not be an appropriate solution to traffic control, particularly as it is associated with a lesser ranking in terms of importance when compared with other police duties. The police have to make day-to-day decisions about allocating their staff resources and it is frequently the case (in the UK at any rate), that insufficient resources are available for checking speeding violations, careless and dangerous driving, and parking offences. Laws and regulations may carry little force if the probability of detection and the perception of detection are so low that they can safely be ignored. In the UK the chances of getting caught in a traffic violation are remarkably small (Dix and Layzell 1983). For drink-driving, they are 1 in 250 but as low as 1 in 2,500 in some areas, for speeding 1 in 7,600, and for illegal parking in central London, 1 in 50. At these levels of detection road traffic laws seem to approach the point at which they have a negligible deterrent effect. A further escape route for offenders exists in the courts, which may have very erratic patterns (over a whole country) of sentencing for road traffic law violations.

LAND USE PLANNING MEASURES

Here we refer to a scale of activity somewhat broader in spatial terms than detailed urban design for residential areas, which is the subject of chapter 7. The increase in length of journey to work, the increased suburb–suburb commuting and shopping trips in cities, and the longer journeys to hospitals and schools are all the result of physical land use planning decisions which are either taken by public bodies or condoned within particular national planning systems, which vary from country to country in Europe.

The development of large regional hospitals, of large out-of-town shopping centres or hypermarkets and of other traffic generators, has proceeded steadily over the years and has been well documented, e.g. in the case of sports facilities (Hillman and Whalley 1977). The increased distances, which have resulted from this process, give considerable advantage to the car user, particularly where the established public transport routes cannot re-adjust very easily to the new demands. Land use planning which is very much circumscribed by the activities of 'free market' forces, as in the case of out-of-town shopping, has not coped very well with these trends and has not consciously pursued an energy and transport-related strategy to improve accessibility and, at the same time, minimize the need to travel. The Commission of the European Communities has referred to the need to develop an optimal spatial distribution of people and activities (EEC 1976). This idea has not developed very well since the mid-1970s, but could produce a land use policy with specific energy-minimizing and travel-minimizing objectives.

PUBLIC TRANSPORT MEASURES

Public transport is a safer mode of transport than walking, cycling, or travelling by car (table 6.10). If passengers are attracted from car to forms of collective transport (bus, metro, S-Bahn, etc.), then there is likely to be a reduction in road accident rates in cities. Similarly, under present conditions of walking and cycling a transfer from these modes to collective transport would produce a reduction in RTAs. The support for public transport varies enormously through Europe (see figure 6.3).

The range of percentage subsidy displayed in figure 6.3 suggests very different attitudes towards public transport and its costs and benefits, but does not provide any direct evidence on the possible relationship between subsidized public transport (or low-fare public transport, which may not be the same thing) and road accident rates.

Table 6.10 Road accident rates, according to different methods of personal travel, 1976

| Method of travel* | Accident rate | | | Fatal/serious accidents | |
	per 100,000 journeys	per 100,000 hours	per 100,000 vehicle-kilometres	number	%
Bus	2.4	4.6	2.9	1,240	2
Walk[†]	11.5	40.2	82.6	20,630	25
Car	17.3	42.3	12.3	34,879	43
Cycle	30.1	111.8	104.2	4,931	6
Two-wheel motor	405.1	568.8	442.1	19,851	24
All methods	17.9	48.3	22.5	81,531	100

*Calculated for annual rates in Department of Transport (1977), table 47.
†Calculated by relating the rate per person to the frequency with which the NTS shows journeys made on foot.
Source: Hillman and Whalley (1979).

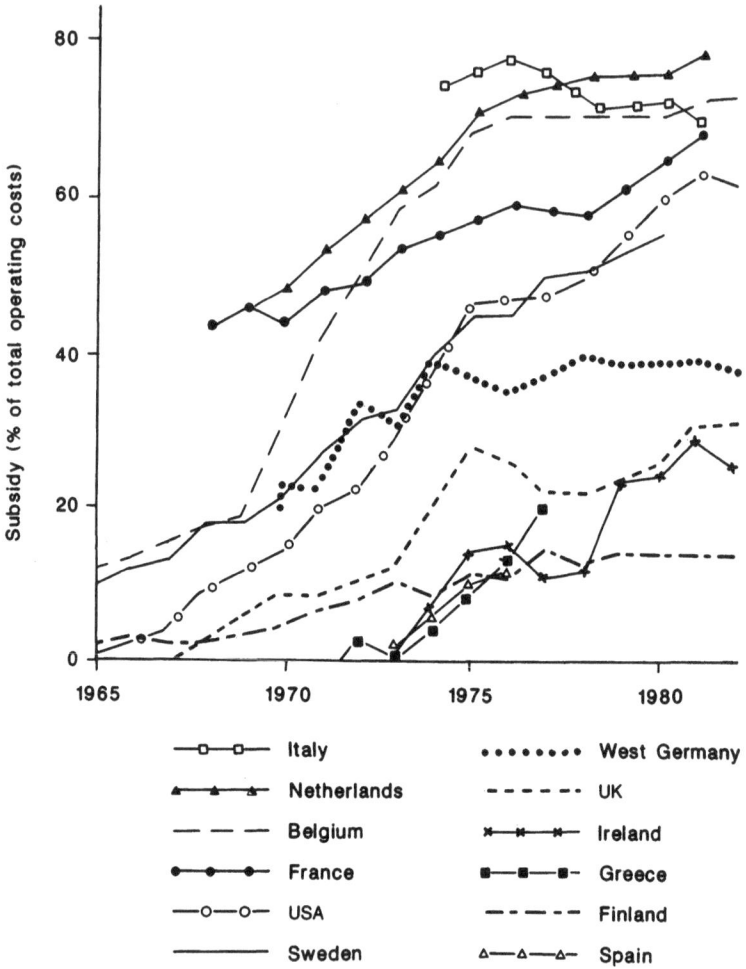

Figure 6.3 Level of subsidy in different public transport systems

In the UK a clear relationship has been demonstrated between public transport use, modal split and RTAs (Allsop 1983). Allsop examined the low-fares policy of a council (Greater London, in operation October 1981 to March 1982) and estimated the excess of casualties in the year May 1982 to April 1983 over the number that would have been expected if the low-fare policy had prevailed and not been ended by a legal action. The results are summarized in table 6.11.

Notwithstanding some heavy qualification of the results by the author, the work demonstrated the importance of public transport support as a means of reducing RTAs. Allsop and Turner give a clear indication of the significance of the results:

> In the low fare period from October 1981 to January 1982 there were 17394 casualties, whilst in the corresponding four month period of high fares prior to the new seatbelt legislation of 31 January 1983, there were 20007. This increase of 15% corresponds to a model estimate of 11% for the whole year. (Allsop and Turner 1984:155)

Similar effects have been noted in Merseyside and the Lothian region of Scotland and are discussed in Whitelegg (1987b).

In those European countries with generally better public transport facilities in cities than the UK, there is not an obvious reduction of accident levels in cities, but there are so many other variables which intrude into the straightforward connection between quality of public transport and accident reduction that much more research would be necessary before general conclusions could be drawn.

In the French study of the town of Mans (Tortrat 1985) it was shown

Table 6.11 Excess casualties as a result of ending low-fares scheme

Class of road user	Estimated percentage of excess casualties
Pedestrian	21.9
Pedal cyclist	11.3
User of powered two-wheeled vehicle	17.7
Occupant of car/taxi	9.9
Occupant of PSV	7.0
Occupant of other vehicle	10.5

Note: The actual number of excess was 6,371 in a higher estimate and 4,292 in a lower estimate.
Source: Allsop (1983).

that a reduction of 26.53 per cent of accidents between 1977 and 1983 was the result of a number of specific policy measures, one of which was 'le development d'une politique en faveur des transports en commun' p. 131.

In the next section we will examine the German road safety programme as an example of one European state's efforts in this area. The German programme (FRG 1984) is wide-ranging in its scope and is designed to give a fresh impetus based on the experience which has accumulated since the 1973 programme was inaugurated. Its general orientation is at variance with the list in table 6.9, in that it concentrates on influencing behaviour: 'The pivotal part of all future road safety work must be the early, but long term, instruction and enlightenment on behaviour in road traffic with a view to developing and strengthening the safety consciousness of road users both for ourselves and for others' (FRG 1984: 9).

The measures to be followed, to achieve improved road safety, fall into a number of distinct categories.

1. The specific improvement of traffic safety for pedestrian children; this involves promoting a 'child and traffic' programme of instruction and direct attention to parents and the necessity for their involvement in this kind of training; there are special programmes for bicycle training, similar to schemes in the UK for cycling proficiency

2. Traffic programmes in schools including cycling, but also traffic education materials overlapping with biology, religion, and 'civics' which is aimed at reducing aggressive behaviour in road traffic

3. Programmes for beginners in motor traffic; for riders of cycles with auxiliary engines ('mofas'), for motor-cyclists, and for young drivers of cars (aged 18-25)

4. Other group-selective programmes, e.g. the lorry safety programme, and the 'elderly persons as pedestrians in road traffic' programme; the Federal government wants to see more emphasis on the elderly in future programmes because of their particular vulnerability

5. General traffic safety information for the public; the objective is to keep the public aware of road safety as an issue and 'focus on eliminating the causes of accidents', e.g. excessive speeding

6. Alcohol - measures to enforce existing legislation more strenuously and to introduce electronic breathalysers; associated measures to alert the public and the medical profession to the dangers of

prescribed medicines and drugs in certain circumstances

7. Speed – the relationship between speed and RTAs is always a matter for great argument and this has been the case in Germany, raising many important issues which are discussed later in this chapter as a special topic; the Federal government has decided on one measure which is of some general interest, and that is it will endeavour to persuade the automobile industry to stop using speed as a way of selling their products

8. Protection for individuals via the use of seat belts in cars and protective helmets for motor-cyclists; the use of seat belts is much lower in Germany than in Britain (45 per cent in urban areas in Germany, compared with compliance levels of over 95 per cent in Britain) in spite of a Federal law which makes belt wearing compulsory; there is a debate about the road safety importance of these matters (Adams 1985) and it cannot be assumed that helmet wearing or belt wearing is a road safety improvement without some damaging side-effects

9. Measures for unprotected road users (cyclists and pedestrians); motorists must be educated in those forms of driving which will protect the road users, particularly prudent driving, reduced speed and improved braking; likewise some measures are needed to curb the 'unruly' behaviour of cyclists

10. Measures to improve the competence of young drivers of motor-cycles, mopeds, and cars; improvements in training and requirements for theoretical and practical tests; the quality of driving instructors should be improved and the driving test itself should be improved

11. Two-stage licences for motor-cycles to try to reduce the very high rates of accidents involving this class of vehicle; the Federal government proposes two classes of licence based on power of the vehicle: (a) unlimited horsepower restricted to persons over 20 years of age, and (b) limited to 27 horsepower which 18- and 19-year-olds will be allowed to ride

12. A Central Road Traffic Register for holding data on breaches of regulations and as a check on fitness to drive

13. Road traffic surveillance is confirmed as a valuable road safety measure and as a way of improving the education of road users

14. Improving vehicle safety including attention to brakes, lighting, helmets, bumpers, and the adoption of EEC regulations in these areas; the Federal government would also like to see the automobile industry develop vehicle design measures which will make less

severe the consequences of accidents involving pedestrians and riders
of two-wheeled vehicles

15. Improving road safety through road construction and traffic control
involves 'closing gaps' in the network of Federal motorways, building bypass roads, eliminating railway level crossings, construction of
cycle tracks along federal highways (2000 km from 1984 to 1990);
and slowing down of traffic in residential areas.

The German example of road safety activity illustrates many traditional
areas of concern to all European countries, but also some novel
approaches. The attention to education, publicity, influencing behaviour,
and the technical standards of vehicles, account for most road safety
activity in European countries, yet it is quite clear that this absolves the
general traffic environment of land uses, activities, and governmental
fiscal support from any direct responsibility (Whitelegg 1983, 1984b).
Moreover, the German programme accepts the benefits of road construction without considering traffic generation effects or other disbenefits
(Transport 2000, 1986). On the positive side, and of great value to road
safety evaluation, is the programme of construction of cycle ways and
growth in the number of areas where pedestrians and cyclists have special
priority.

Speed limits

There is a relationship between driving speeds and the occurrence of
accidents. This was the theme of more than one submission to the House
of Commons Select Committee on Transport during its investigation of
Road Traffic Accidents (House of Commons 1985).

Lower speeds, of course, are going to be more in harmony with other
road users, particularly in urban areas. Lower speeds give more time for
the driver to detect problems, consider the course of action to take, make
the decision, and carry it out. Braking distances are reduced, conferring
an immediate road safety benefit. If accidents do happen, the result of
mass times velocity is more in favour of reduced severity of accidents. Lower
speeds give pedestrians and cyclists more time to take avoiding action,
and to gauge their behaviour in relation to general traffic speeds with
some expectation that there is order and certainty in traffic situations,
rather than chaos and the unexpected. Mutual expectations with a high
degree of reliability are a better guarantee of behaviour which minimizes
accidents than is uncertainty.

Table 6.12 Relationship between reduction of median speed and reduction of accidents with different severity (per cent)

V_1 – initial speed (km/h)
V_2 – reduced speed (km/h)

Reduction of fatal accidents $\qquad 1 - \left[\dfrac{V_2}{V_1}\right]^4$

Reduction of injury accidents $\qquad 1 - \left[\dfrac{V_2}{V_1}\right]^3$

Reduction of accidents reported to police $\qquad 1 - \left[\dfrac{V_2}{V_1}\right]^2$

Source: Nilsson (1981).

Speed has other deleterious side-effects. It increases noise, it raises the engineering standards for width and alignment, so making the road/car environment more dominant to the detriment of other users. It reduces visual order as more and bigger signs to control and advise traffic respond to higher speeds. It causes territorial intrusions and discomfort to residents, it increases perceived risk and fear, and it increases actual risk of accidents and the severity of these accidents which are caused.

Evidence on the effects of speed is not in short supply (OECD 1981; Haworth 1985; Nilsson 1981), but as in many aspects of motoring, traffic law, and behaviour, the whole issue is clouded by myth and supposition and home-spun philosophies, which weave a pseudo-scientific system of evidence for the safety benefits of speeds above the generally prevailing levels. The mythology takes root in a fertile ground of police inactivity and tacit support, and in convoluted governmental statements proclaiming various versions of the freedom of the individual. The evidence of police representations to the House of Commons Select Committee (House of Commons 1985) are revealing in the former respect.

Nilsson's (1981) evidence is particularly illuminating. Based on extensive research in Sweden, Nilsson produces generalized results on the effects of speed (see table 6.12). This is reinforced by Rumar (1985) in his graphic illustration of the relationship between accident and speed (figure 6.4).

On the governmental viewpoint, the German Federal Ministry of Transport is particularly good at mental gymnastics:

> The Federal Government does not believe that a general reduction of the speed limit in built-up areas (to 30 km/hr) is likely to promise any success. The fact of an existing regulation does not necessarily elicit responsive behaviour on the part of the driving community. What really counts is a 'social consensus' as a rule. And such a consensus ... certainly does not exist. (FRG 1984:13)

The Federal government prefers to believe that accidents are not attributable so much to high permissible speeds, 'but rather to the inappropriate conduct of individual drivers in a given traffic situation'. This is

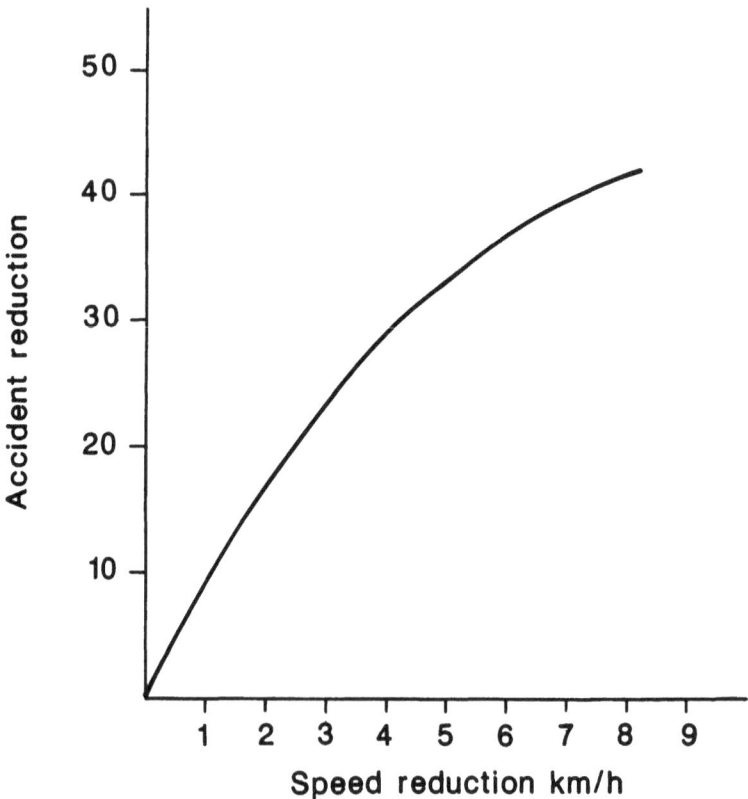

Figure 6.4 Relationship between accident reduction and speed reduction

Table 6.13 European speed limits (in kilometres per hour)

Country	Built-up areas	Outside built-up areas	Motorways
Belgium	60	90	120
Denmark	60	80	100
France	60/80*	90/110*	110/130*
West Germany	50	100/130*	130 (recc.)
Greece	50	80	100
Ireland	48	64/88*	–
Italy	50	80/110*	90/140*
Luxembourg	60	90	120
Netherlands	50	80	100
Portugal	60	90	120
Spain	60	90/100	120
UK	48	100	112

*For details of restrictions on classes of vehicles, length of driving experience, and whether touring or not, see original source.
Source: RAC Continental Handbook and Hotel Guide, 1986.

really fudging the issue, but is a good representative view of more than one governmental position in Europe. The general position in Europe (in selected countries) as regards speed limits, is shown in table 6.13.

West Germany is the only European nation without a speed limit on its motorways, and it is in that country where the debate about speed limits has been particularly fierce. Speeds on the motorways have increased steadily in the past thirty years (see figure 6.5), apart from a period of a few months in the energy crisis of the early 1970s when a 100 k.p.h. limit was enforced. The arguments for a speed restriction on German motorways have concentrated on environmental issues (see chapter 11), particularly on acid rain and damage to forests. Germany is heavily forested (28 per cent of its land surface) and motorways make frequent use of forests for their routeways. The arguments for the speed limit are, however, much more wide-ranging than environmental, and Retzko and Sturm (1986) have reviewed the effects of speed limits on a large number of variables, including the automobile industry, energy, and road safety. The automobile industry has responded vigorously with a campaign against the speed limits (VDA 1985).

The German experience shows the amount of disagreement on this basic transport topic. Road traffic accident statistics can be used to

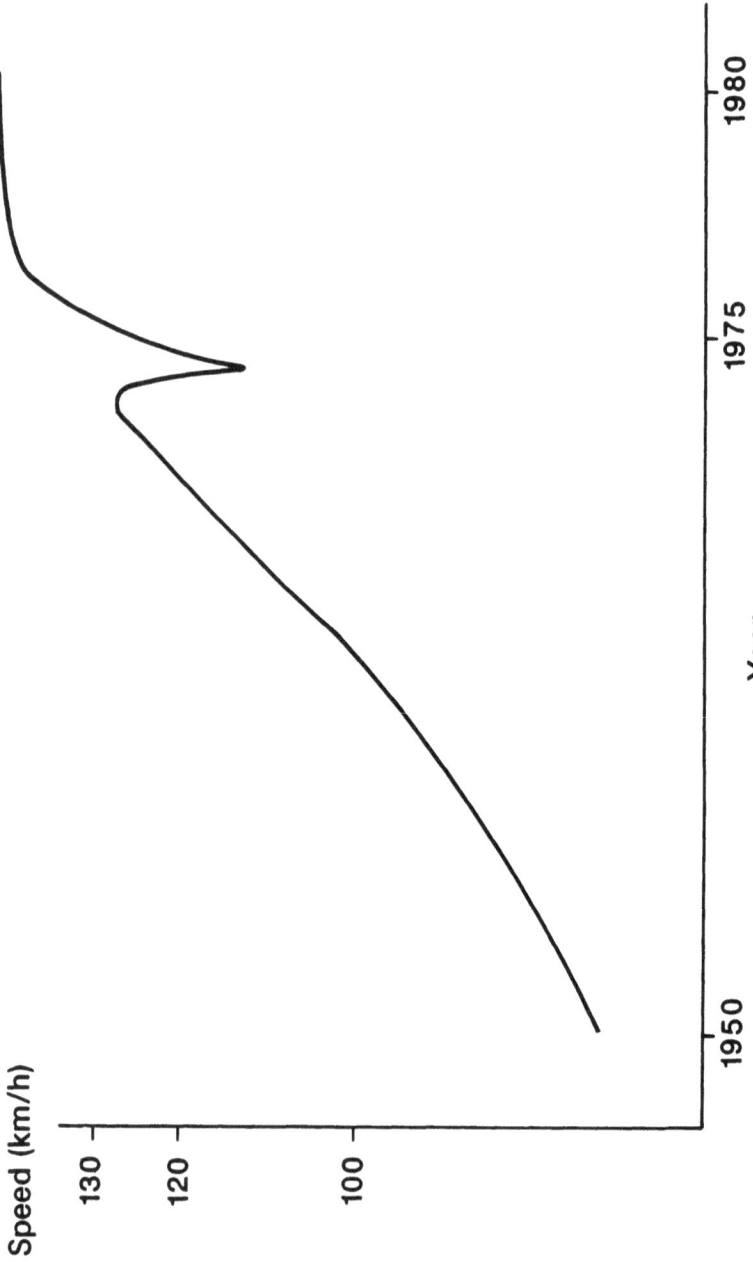

Figure 6.5 Development of speed on German motorways

support both sides of an argument on speed restrictions, and there is a wide measure of professional disagreement and strong pressure-group lobbying. The German government has opted for the status quo and is heavily influenced by the stance of the automobile industry. The decline of death and injury in most European countries provides a ready excuse for inaction, as does the relative safety of motorways compared with other roads. These positions fail to take sufficient notice of the number of accidents themselves and of the redistribution effect on unprotected road users. There is also very little research on the type of serious injuries which are sustained in RTAs and whether these have changed over time. Quick response rates on the part of emergency services and sophisticated hospital care might well produce a reduction in deaths that is nothing to do with the accident situation itself. Similarly, hospital care is capable of extending life beyond the 30-day limit prescribed in several definitions of an RTA fatality. The decline in RTA fatalities and injuries, as an aggregate statistical tendency, is not a reliable indication of improvements in road safety and much more attention needs to be paid to the motorized environment of RTAs and their changing circumstances over time.

7 Pedestrians, cyclists, and urban design

Pedestrians

In most countries there is a long-standing and well-established tradition of neglecting the pedestrian. This is not a matter simply of neglect, as other transport and planning policies frequently make walking very difficult, and occasionally impossible, and so represent policies which conspire to eliminate walking as an important mode of transport. Policies which advocate urban road construction, underpasses, out-of-town shopping centres, and large centralized facilities of any kind which are remote from public transport routes are not 'pedestrian-friendly'. Policies which make the streets more dangerous, or fail to tackle the problem of traffic intruding into residential areas, heighten a sense of fear and insecurity, suppressing pedestrian trips. Walking is highly sensitive to environmental factors whether these be meteorological, topological, or psychological; adverse reactions will be heightened among the elderly, the very young, and those in charge of the very young, and among those groups threatened with personal violence on their walking journeys through brutal and ill-designed urban environments. The situation for women is particularly bleak, as they balance the psychological disadvantages of isolation and incarceration against the real risk of sexual violence on public transport and in urban streets (Oliver 1986). Those in charge of children experience similar problems when contemplating the growing demand for independence and mobility from their maturing offspring.

In spite of real difficulties with the walking environment, walking as a mode of transport is very important indeed. In a major study of walking, Hillman and Whalley (1979) were able to show that about one journey per person per day was made on foot, and over one in three of all journeys that people made were walking journeys: 'This is only a slightly lower proportion than that for car travel, and far higher than for all other

Table 7.1 Modal split on all journeys

	Walk	Car	Bus/train	Cycle	Other	n/a	All journeys*
All journeys (%)	35.2	39.5	11.3	3.2	1.2	9.6	100
Trip rate- journeys per person per day	0.9	1.01	0.29	0.08	0.03	0.24	2.56

*Total number of journeys, 71,474.
Source: Hillman and Whalley (1979), table II.1, p. 20.

methods of travel, including public transport. Indeed it amounts to over 48 m. journeys being made on foot each day in Britain' (Hillman and Whalley 1979: 19).

This study excludes walking journeys of less than 50 yards and excludes journeys on foot of the kind which involve other modes (e.g. a walking journey to a bus stop or a walk from car park to office or shops). The real extent of walking is, therefore, going to be much greater than published statistics show. Table 7.1 reproduces some modal split results which show the relative importance of different modes of transport usually measured in percentages of trips which are made on foot, by car, by bus or train, etc.

Walking is more common for some journey purposes than for others. A high proportion of journeys to and from school and shops are made on foot, but the proportion for work journeys is low. Since some journeys are made more frequently than others, the importance of walking varies in response to its share of particular purposes (see figure 7.1).

Shopping accounts for nearly one-tenth of all personal transport and is a particularly important destination for travellers on foot. In a survey of travel to a supermarket in north London, 42 per cent of shoppers arrived on foot, compared with 50 per cent by car and 6–7 per cent by bus. It is this close connection between shopping and walking that makes the design of pedestrian precincts and pedestrianized streets an important feature of urban areas and the satisfaction of pedestrian needs (London Borough of Brent 1986).

A study in London (TEST 1985) showed that walking accounted for 36.3 per cent of all trips and 16 per cent of work trips. This figure is in line with a selection of German cities reviewed in the German 'Future of the Automobile' project (Jeschke and Kunert 1984). This is summarized

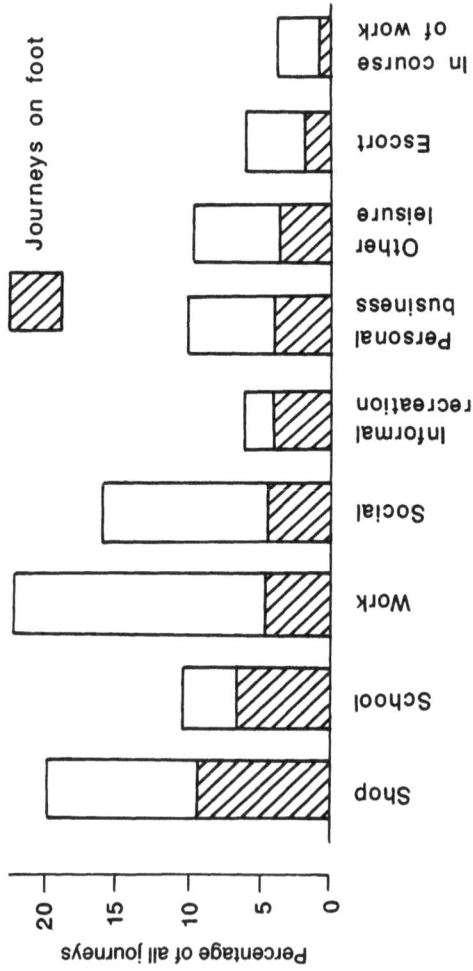

Figure 7.1 Characteristics of walking journeys

in table 7.2. In German cities the percentage share of trips for walking varied between 20 and 40, with Bielefeld having the highest percentage share and Erlangen the lowest.

The provision of facilities for pedestrians can take many forms all of which are ultimately bound up with good urban design, traffic restraint, and high-quality physical environments in cities. Eliminating or reducing traffic significantly will lead to a dramatic improvement in conditions for the pedestrian. This is the basic idea behind pedestrian streets and pedestrian malls and areas, and ultimately perhaps of the central areas of cities which are traffic-free or very much less trafficked than at present. Providing facilities for pedestrians must also embrace the humble objects of design such as pavements, kerb heights, and obstructive street furniture. The design of attractive walkways, a 'human-scale' of architectural and landscape interest, the absence of noise and pollution, together with a sense of security, safety, and well-being, will all encourage walking activities as well as producing an urban environment which is attractive alike to workers, businesses, and tourists. It is in this wider sense that pedestrian facilities are so important, and their neglect a cause for concern in Europe, where many cities are suffering physical and economic decline.

This broader view is well developed by TEST (1985), in a study which focuses on the City of London and its physical environment, which is much degraded by heavy traffic and poor facilities for pedestrians. While the City of London is rather unusual in its constitution, global importance as a financial centre, and grossly inadequate facilities for pedestrians, it does present us clearly with the problems of planning for walking journeys. The City's resident population was 5,900 in 1981, but its working population was 299,000. The day-time population is further boosted by tourists; in 1983, St Paul's Cathedral had 2.5 million visitors (TEST 1985). Pedestrian activity within the City is very high; in 1977, 64 per cent of commuters arriving at Liverpool Street/Broad Street BR stations, and 84 per cent of those at Cannon Street, completed their journey on foot.

High pedestrian flows in the City contribute to a high accident toll. Table 7.3 shows that in London – which itself has a high accident rate compared with the rest of Britain – the City of London comes out very badly indeed.

TEST (1985: 41) reinforces the obvious conclusion that life is particularly difficult for pedestrians: 'It is remarkable how closely [the accident rates] correspond to activity patterns, with the two cities with the highest

Table 7.2 Modal split information for German cities

City or metropolitan area	Year of survey	Population	Trips/person gross*	net†	Trip maker ratio	Modal split walk	bicycle, moped	motorcycle, passenger cars (driver and passenger)	Transit
Hamburg	1976	2.8m.	(1.21)	–	–	–	0.13	0.60	0.48
Munich	1970	2.1m.	(1.13)	–	–	–	–	0.78	0.35
Berlin (West)	1976	2.0m.	2.4 (100%)	–	–	0.75 (31%)	0.07 (3%)	0.8 (33%)	0.8 (33%)
Nuremberg	1975	1.2m.	2.57 (100%)	–	–	0.8 (31%)	0.15 (6%)	1.18 (46%)	0.44 (17%)
Dortmund	1974	626,000	2.34 (100%)	–	–	0.8 (33%)	0.03 (2%)	1.03 (43%)	0.48 (20%)
Bremen	1976	562,000	2.73 (100%)	–	–	0.70 (26%)	0.32 (12%)	1.17 (43%)	0.54 (20%)
Duisburg	1970	455,000	(1.14)	–	–	–	0.13	0.62	0.39
Unna(Kreis)	1977	390,000	2.0	–	–	0.61 (30%)	0.19 (10%)	1.00 (50%)	0.19 (10%)
Bielefeld	1969	370,000	2.34 (100%)	–	–	1.0 (40%)	0.21 (9%)	0.80 (36%)	0.33 (15%)
Freiburg	1976	176,000	3.36 (100%)	–	–	1.02 (30%)	0.39 (12%)	1.46 (43%)	0.49 (15%)

	Year	Population						(1)	(2)	(3)	
Leverkusen	1975	170,000	2.55 (100%)	–	–	0.64 (25%)	0.34 (13%)		1.20 (47%)		0.37 (15%)
Salzgitter	1974	122,000	(2.18)	–	–	–	0.16		1.72		0.30
Bottrop	1976	117,000	(1.80)	–	–	–	0.30		1.27		0.23
Erlangen	1974	100,000	3.16 (100%)	–	–	0.96 (20%)	0.41 (13%)		1.39 (44%)		0.40 (13%)
Neustadt (Weinstraße)	1977	55,000	3.50 (100%)	–	–	1.07 (31%)	0.31 (9%)		1.89 (53%)		0.23 (7%)
Sindelfingen	?	54,000	3.40 (100%)	–	–	1.03 (30%)	0.30 (9%)		1.77 (52%)		0.30 (9%)
Size of town		–2,000	2.17	3.31	0.66	24.1%	10.3%	1.1%	41.1%	11.6%	11.5%
		–5,000	2.18	3.32	0.66	30.6%	9.1%	1.1%	38.3%	10.8%	9.9%
		–20,000	2.45	3.42	0.72	29.5%	11.8%	0.9%	38.4%	10.9%	8.2%
		–100,000	2.52	3.54	0.71	31.1%	11.4%	0.9%	36.2%	11.3%	8.7%
		–300,000	2.52	3.49	0.72	29.5%	9.1%	0.9%	37.4%	10.1%	12.8%
		–500,000	2.34	3.33	0.70	33.4%	2.5%	0.3%	37.0%	11.5%	15.2%
		500,000–	2.48	3.38	0.73	32.9%	4.7%	0.5%	32.8%	9.4%	19.4%
Average			2.42	3.43	0.71	30.3%	9.6%	0.9%	36.9%	10.7%	11.4%

*All persons over the age of 10 years.
†All trip makers over the age of 10.
(1) – motorcycle
(2) – car driver
(3) – car passenger

pedestrian densities ... leading the list' (see table 7.3). TEST (1988) develops the thesis that good pedestrian facilities are closely equated with high-quality environments and that this, in turn, is a direct stimulus to economic improvements. In support of this argument several different pieces of information are pulled together:

1. Retail turnover increases after a street has been pedestrianized
2. London's Covent Garden has experienced land value rises since major environmental improvements; rents have increased by 50 per cent in a three-year period
3. In downtown Boston (USA), area-wide traffic restraint has led 80 per cent of businesses to experience increased sales; rents for retail space rose by 40 per cent in the 1978–80 period, immediately after scheme implementation
4. A summary of businesses in Hackney revealed that seven out of thirty-three firms intended to leave the borough primarily for environmental reasons.

TEST admit that the evidence is partial and that most of it can be qualified, but they do raise the very important and very clear issue of the economic benefits of improved environment, urban design, and pedestrian facilities. European cities provide widely varying circumstances in terms of pedestrian facilities, particularly of streets converted for pedestrian priority. Table 7.4 gives comparative data from several different European cities.

Pedestrian streets and pedestrian areas have a long history in Europe. The first was introduced, in 1929, in Essen, Germany, and the first in Britain, in 1967, in London Street, Norwich. These dates relate to conversions of previously trafficked streets. In most European countries there are examples of covered malls and shopping arcades that go back over 100 years, and in Britain there are post-1945 New Town schemes which provided such facilities as part of the original design. Roberts (1981), in a definitive study of pedestrian precincts in Britain, estimated that there were 1,450 units (where a unit is either a single converted shopping street, or an arcade, or an enclosed centre).

The rise in the number of converted streets in the USA, West Germany, and the UK is shown in figure 7.2. The figure shows an increasingly rapid development from about 1960 in West Germany, and 1967 in the UK and 1965 in the USA and Canada. It is now common-place to see such pedestrian zones in European cities, as well as their

Table 7.3 Number of road traffic accidents involving death or injury to pedestrians and vehicle occupants per kilometre of borough road, 1977

Borough	Injury accidents per kilometre of borough road
Inner London:	
Westminster	11.64
City of London	10.06
Kensington & Chelsea	8.80
Camden	7.42
Lambeth	6.73
Islington	6.53
Hammersmith & Fulham	6.03
Outer London:	
Hounslow	4.43
Greenwich	4.39
Waltham Forest	4.18
Ealing	4.04
Brent	4.03
Hackney	5.89
Southwark	5.62
Wandsworth	5.39
Tower Hamlets	4.93
Lewisham	4.60
Haringey	4.12
Newham	3.96
Harrow	2.27
Greater London	4.10

Source: TEST (1985).

Table 7.4 Pedestrian facilities in European cities

European city	Length of street (km)	km/km²
Munich	3 + market area	1.96
Vienna	2.35	1.53
Leeds	1.7	1.1
Central London	1.1	0.04

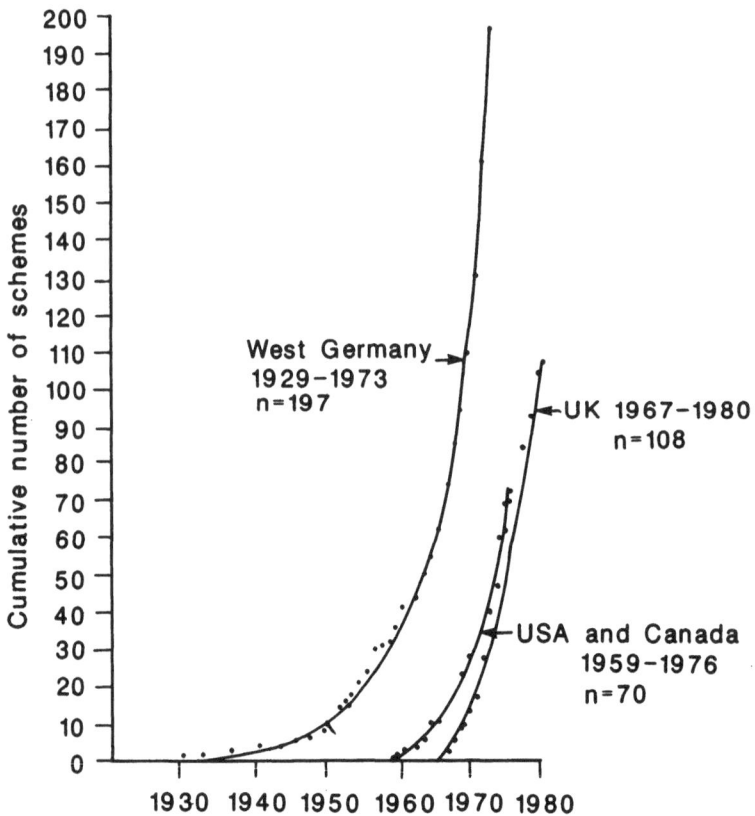

Figure 7.2 Numbers of streets converted for pedestrian use

negative associations which range from concentrations of traffic around the zone, making penetration very difficult, to the neglect of the non-pedestrianized zones in cities. In many European cities residential areas near city centres suffer greatly from the ingress of traffic as a consequence of freeing the protected zone, and the condition of buildings and historic features also suffers from the operation of a dual standard of conservation. The selection and isolation of pedestrian zones as 'special' areas may well be seen as counter productive in terms of wider traffic and residential area concerns, especially if they have emerged independently of a general traffic and public transport strategy.

Facilities for pedestrians which embrace many, if not all, of the ideas and objectives to be found in TEST (1985) are now much more common in Europe than ten years ago. In the Netherlands and West Germany there has been considerable progress in the development of residential areas, with considerable traffic restraint and urban design features which emphasize the priority given to pedestrians and cyclists. In the Netherlands, the *Woonerf* is a well-established feature of urban areas and provides the model for the West German *Verkehrsberuhigte Bereiche*, or traffic restraint precinct. In a survey conducted over late 1983 and early 1984, 541 such precincts were identified in the Federal Republic (Bundesminister für Verkehr 1987). These areas, sometimes known as the '325 areas' after the relevant number in the German highway regulations, have the following characteristics:

1. Pedestrians may use the entire width of the street; children are allowed to play everywhere
2. Vehicular traffic must proceed at walking pace
3. Motorists may neither endanger nor impair pedestrians; they must wait if necessary
4. Pedestrians may not unnecessarily impair vehicular traffic
5. Parking outside the specially marked areas is not permitted except for picking up or setting down passengers, loading, or unloading.

Janssen (1985) describes some elements of Dutch Woonerven schemes in Rijswijk and Eindhoven in the 1979–81 period. In each of these towns the road network of a 'self contained urban area of some 100 hectares' was reclassified and reconstructed to achieve certain objectives:

1. Separate traffic zones from residential zones
2. Within traffic zones, separate different kinds of traffic and to improve the safety of pedestrians and cyclists
3. Within the residential areas (a) to keep out through traffic, (b) reduce the speed of local traffic, and (c) make the environment attractive.

Within the residential zones, a careful 'before and after' analysis was conducted which revealed a dramatic reduction in the number of accidents involving an injury. The actual number of accidents involving injury in these zones was 50-60 per cent lower than expected, on the basis of developments in certain control areas. This is illustrated in figure 7.3.

This favourable effect in residential areas is attributed to the reduction

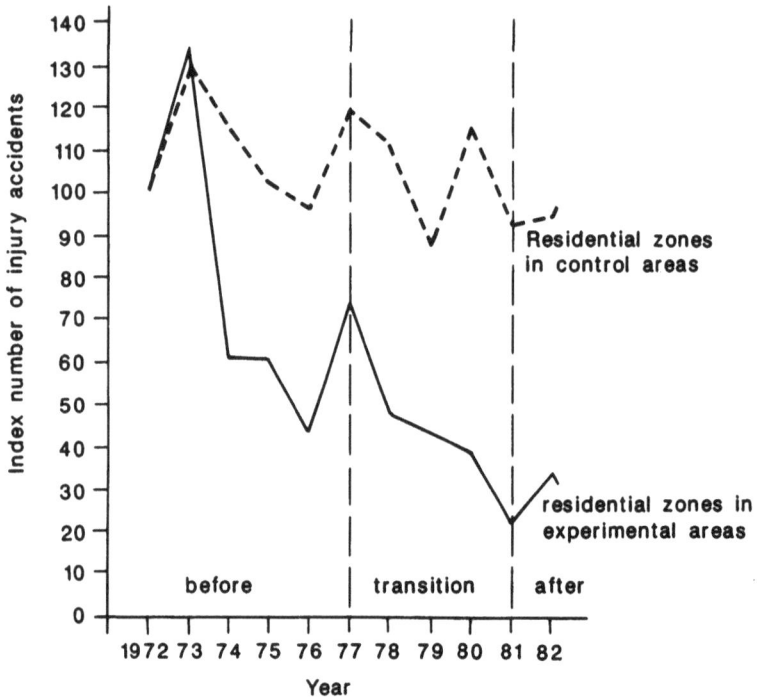

Figure 7.3 Number of injury accidents in residential areas before and after Traffic Quietening

in the amount of through traffic, which would also produce significant benefits in reduction of noise and general improvements in quality of life, particularly social interaction and neighbourliness as described by Appleyard (1981), and in German residential areas in Dusseldorf (North Rhine Westphalia 1979). Figure 7.4. shows the results of such traffic reduction on social interaction.

Following street improvements of the kind shown in figure 7.4, pedestrians are liberated from the constraint of a narrow pavement width and the possibility of only a very few street crossing movements due to traffic volumes. The same idea, which can be most beneficial for single streets, is applicable to larger areas and particularly so in older residential areas of cities, where some protection from through traffic and non-residential parking can produce remarkable improvements in quality of life for residents.

PEDESTRIANS, CYCLISTS, AND URBAN DESIGN   117

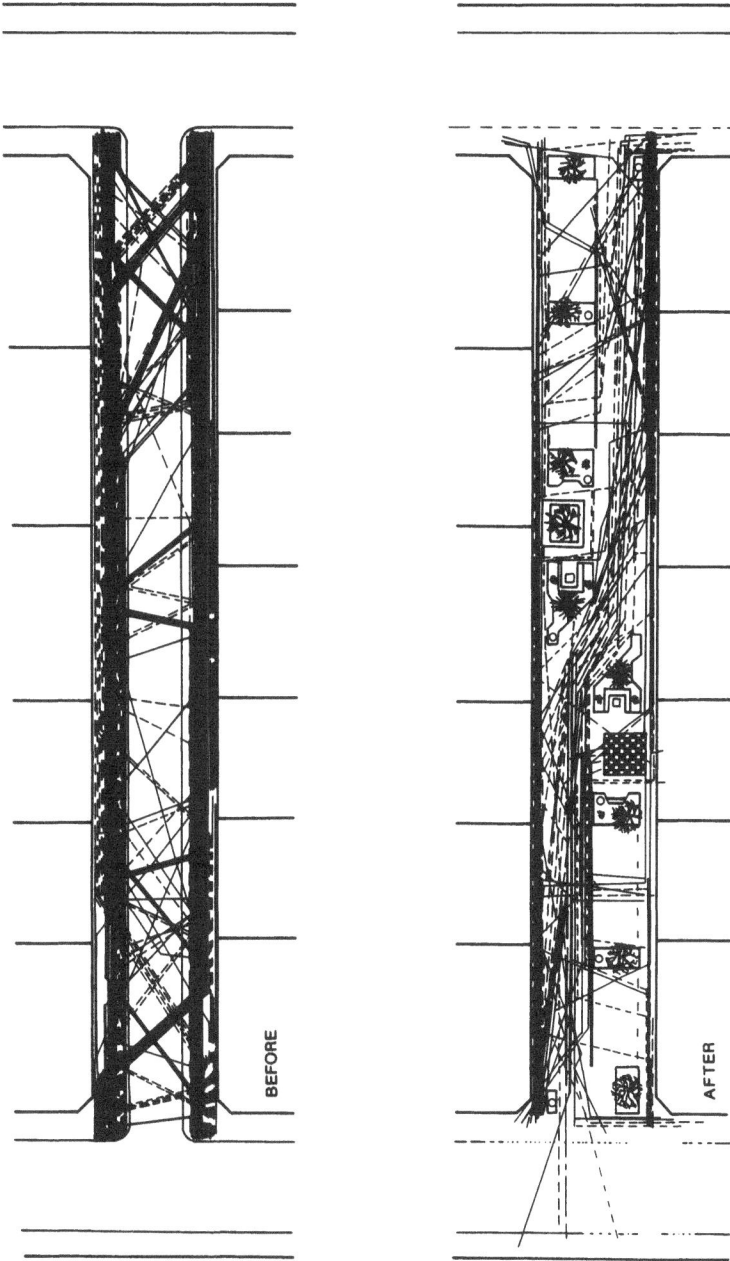

Figure 7.4 Effects of redesigning an urban street on patterns of pedestrian movement (Gelsenkirchen)

Figure 7.5 Example of street design under 'Verkehrsberuhigung': before and after (Berlin)

Figure 7.6 Example of street design under 'Verkehrsberuhigung': before and after (Ortskern Wollmatingen Nord)

Appleyard (1981) has shown how the traffic density along residential streets regulates the frequency of personal contact, and hence friendship and acquaintance patterns. His studies in San Francisco, California, showed that on lightly trafficked streets (2,000 vehicles per day), the residents were found to have three times as many local friends and twice as many acquaintances as those on a 'heavy' street (16,000 vehicles per day). On heavy streets, there was very little social interaction. Residents had few, if any, friends there.

The German and Dutch experience of street design and residential area wide improvement leads the field in Europe. Figures 7.5 and 7.6 show, in sketch form, exactly what is done in these policies; and both show a 'before' and 'after' view. The benefits of such reduction in traffic volumes, and associated reduction in noise, pollution, and stress, are well documented by Appleyard (1981) and lead very quickly to dramatic improvements in the quality of life for residents. Area-wide schemes, as in Chalottenburg in Berlin and Eberstadt in Darmstadt, take the concept to a higher geographical scale but can produce traffic problems on the edges. These 'rings of steel' blight many historic city cores by encircling them with very high volume traffic flows (e.g. Lancaster and Cologne).

Europe, with its high population densities, long tradition of urban living, and rich legacy of historic urban forms and cultural artefacts, is particularly vulnerable to damage from motorized traffic and its associated pollution. Pedestrian activity is one of the first sufferers from this motorization, and the effect is to divorce human activities from their traditional place and scale of operation. Walking is appropriate, if not symbiotically interlinked, with urban civilization. Pedestrian schemes, whether in city centres or residential areas, can go some way to undoing the damage brought about by motorized traffic, but only within a much broader strategic framework for metropolitan planning, land use, and public transport. In most European cities this is poorly developed, and in the UK it has been deliberately destroyed.

Walking and the needs of the pedestrian have been completely neglected by the EEC and its transport policies. The lead which could have been given for a problem which possesses a genuine European dimension has not emerged with serious consequences for the quality of life of European citizens.

In the next section, cycling is examined as another example of a mode of transport with considerable potential for investment; and improvement of benefit to users and society as a whole.

Cycling

Cycling as a mode of transport is on the increase in Europe (see figure 7.7 for a picture of cycling in one large European city), and with this tendency comes a very real road traffic accident hazard. Cyclists, like pedestrians, are very vulnerable in traffic. In 1980, in twelve European countries, 3,892 cyclists were killed and 132,077 injured. The share of cyclists killed ranged from 2 to 21 per cent of all fatalities. In five countries this share was above 10 per cent, in four others it was above 5 per cent.

Cycling has many advantages to set against the disadvantage of vulnerability. It is also a mode of transport which can be catered for by physical improvements to facilities in a way which will reduce accidents.

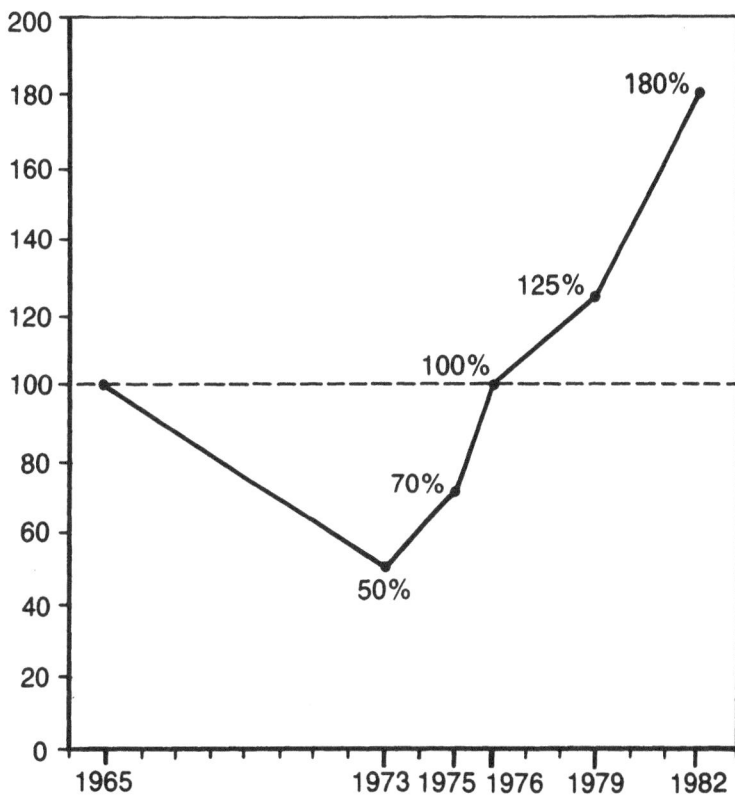

Figure 7.7 Development of cycling in West Berlin, 1965–82.

Source: Der Senator für Stadtentwicklung und Umweltschutz, Berlin, September 1983.

Table 7.5 Cycling policies in Europe

	Austria	Belgium	Denmark	Finland	France	FRG	Ireland
The use of the bicycle is encouraged	Yes	Yes, not governmental	Yes	Yes	Not encouraged not discouraged	Yes	
The reasons for this	Healthy; environment; energy saving					Energy saving; environment healthy	
Concrete measures to encourage bicycle use	Construction of bicycle facilities		To create safe bicycle facilities	To create safe bicycle facilities	To create safe bicycle facilities	Construction of cycle tracks; allowing cycles on field paths; encouraging the federal states to take additional measures, research projects	
Promoting bicycle safety is focused on:							
(a) the rider							
training courses for children	−	+	+	−	+	+	
safety campaigns	+	+	+	+	−	+	
traffic rules	+	−	+	+	+	+	
(b) the bicycle							
visibility	+	+	+	+	+	+	
construction standards	+	+	+	+	−	+	
inspection campaigns	−	−	+	−	−	+	
(c) the road							
segregation of vehicles	+	+	+	+	−	+	
experiments for the construction of cycle routes	−	−	+	+	−	+	
programmes for the construction of cycle paths	+	−	+	+	−	+	
preparing of guidelines	−	+	+	−	+	+	
There are special funds for: the construction of '							
bicycle facilities	not known	+	−		+	+	
safety campaigns	not known	+	+		−	+	

Source: ECMT (1986b).

The arguments for supporting cycling as a mode of transport are powerful. Cycling has clear advantages in energy consumption and pollution, it
is cheap and widely available, it can make a contribution to reducing
congestion in cities, it is healthy for the user and those with whom it

Italy	Netherlands	Portugal	Spain	Sweden	Switzerland	UK	Norway
Yes	Yes	Not encouraged, not discouraged	Not much; not governmental	Yes	Yes, private organizations	Yes, indirect	Yes
To reduce congestion; energy saving	Energy saving; to reduce congestion, environment		There is little cycle traffic in Spain		Environment; energy saving; to reduce car traffic in towns, construction of cycle paths and cycle routes	To reduce congestion, energy costs; pollution; healthy, cheap, efficient	Environment, healthy
Construction of bicycle facilities; parking facilities	To create safe bicycle facilities; parking facilities; construction of cycle tracks			Construction of bicycle facilities			Construction of bicycle facilities
–	+	+	–	–	+	+	+
–	+	–	+	+	+	+	+
–	–	–	+	–	+	–	+
–	+	–	+	+	+	+	+
–	–	–	–	–	+	+	–
–	–	–	–	–	sometimes	–	+
+	+	–	–	+	+	+	+
+	+	–	–	–	+	+	+
–	+	–	–	+	+	+	+
–	+	–	–	–	–	–	+
local	+	–	–	+	–	+	+
+	+	–	–	+	–	+	+

might come into 'conflict', and it is remarkably efficient in cities in providing better access to a range of facilities over distances of up to 5 km than either car or bus. In the role of a cheap vehicle in cities the bicycle can make an enormous contribution to the quality of life of cyclists and non–cyclists alike, and contribute to the regeneration of city cores now in the grip of motorized traffic.

Europe has provided some of the best examples of planning for bicycle transport (particularly in Denmark and the Netherlands), yet it still lacks the support which would ensure its emergence as a major mode of transport in urban areas. Table 7.5 summarizes policies concerning the bicycle in fifteen European countries.

The table is slightly misleading in one important respect, in that most countries claim to encourage the use of the bicycle when this is not the case. Bicycling is not aided by tacit support, as in the UK, but needs a positive commitment in terms of the construction of cycling facilities and in terms of the integration of bicycle planning into an attack on road safety. The construction of new roads, roundabouts, and complex junctions, without the provision of specific cycling facilities, will harm the interests of the cyclist.

Cycling is a difficult activity to measure; there are several reasons for this, including its traditional treatment as an insignificant mode of transport which has not encouraged much activity in the statistics of cycling, unlike car ownership and use. In addition, there may well be a marked divergence between cycle ownership and use, as individuals make difficult calculations about the risk of road traffic accidents and the complicating factor of weather. Nevertheless, cycle ownership is high and increasing; table 7.6. shows this information.

The table shows a remarkable growth in bicycle ownership and equally remarkable geographical differences. The British bicycle ownership level is one quarter that of the Dutch, and while topography and

Table 7.6 Number of bicycles per 1,000 inhabitants, 1973–4 and 1980

Country	1973–4	1980
Netherlands	550	738
Sweden	491	697
Denmark	590	675
Norway	448	489
Belgium	296	339
Switzerland	236	306
France	216	281
UK	130	180

Source: ECMT (1986b).

Table 7.7 Bicycle production data

Country	×1,000 production	×1,000 exports	×1,000 imports	×1,000 bicycles on the road	Year	Population (millions)	×1,000 domestic sales
GB	1,244	270	540		1985	56.4	1,514
NL	876	261	267	11,000	1985	14.3	872
Belgium	240	14	278	3,563	1983	9.9	503
Italy	2,400	1,415	25		1985	56.7	1,009
France	1,635	516	890		1985	54.3	2,009
FRG	2,891	1,045	425		1985	61.4	2,270
Denmark	244				1983	5.1	
Portugal	94				1982	10.1	
Spain	965	80	40		1983	38.4	955
Greece	14				1978	9.9	200
Ireland						3.5	
Luxembourg						0.4	

Source: European Parliament (1986).

Table 7.8 Cyclists' travel behaviour, UK (percentages)

Incidence	
Daily	57
2–3 times per week	30
Once per week	6
Less frequently	7
Journey length	
0–2 miles	8
2–5 miles	25
5+ miles	67
Annual mileage (1980)	
0– 500	15
500–1,000	17
1,000–1,500	14
1,500–2,000	12
2,000+	42

Source: Watkins (1984).

possibly even the weather could be invoked as partial explanations, the main reasons for such a difference lie in the area of bicycle planning and the provision of facilities which make cycling easier and safer. A further indication of the importance of the bicycle in Europe can be given by production data; this is shown in table 7.7. Domestic sales are highest in the Federal Republic of Germany and remarkably high in the UK, given the lack of facilities in the latter country when compared with the Federal Republic, Denmark, and the Netherlands.

Surveys of actual use of bicycles are relatively rare. One such survey (Watkins 1984) provides some interesting tabulations. The survey was carried out among 30,000 members of the Cyclists' Touring Club (UK) and produced a response of approximately 9,000. This sample, biased towards 'committed' and leisure cyclists, probably underestimates the amount of short-journey utility cycling in urban areas. Nevertheless, it provides valuable insights into the travel behaviour of the cyclist. Table 7.8 summarizes incidence of cycling, journey length, and annual mileage; it should be compared with table 7.9, which shows the frequency distribution of cycle-trip distances in a German study.

The National Travel Survey (NTS) (Department of Transport 1983) is based on a sample of 10,000 households and is a more representative

Table 7.9 Frequency distribution of trip distances by bicycle

Distance (km)	Absolute	%	Cumulative %
Up to 1	2,581	35.9	35.9
1–2	2,165	30.12	66.02
2–3	1,102	15.33	81.35
3–4	523	7.28	88.63
4–5	285	3.96	92.59
5–6	148	2.06	94.65
6–7	117	1.63	96.28
7–8	84	1.17	97.45
> 8	184	2.55	100.00
Totals	7,189		

Source: H. Hautzinger and P. Kessel, Mobilitat im Personenverkehr Forschungsauftrag der Prognos AG, Basel, Durchgefuhrt im Auftrag des Bundesministers für Verkehr, Bonn, Forschung Strassenbau und Strassenverkehrstechnik, Heft 231, Bonn, 1977.

national sample for deriving patterns of cycle ownership and use (though it is based on 1978–9 data and only picks up information on 3.6 per cent of all 'journey stages'). However, the NTS data are useful for describing journey purpose (see table 7.10).

The journey to work accounts for the biggest share of journey purposes and for the highest mileages. For a balanced picture of cycle use,

Table 7.10 National Travel Survey data, journey purpose by bicycle (percentages)

Journey purpose	Distribution of bicycle stages	Distribution of distance cycled
To/from work	31	35
In course of work	8	7
Education	11	10
Shopping	16	11
Personal business	9	8
Entertainment and sport	6	8
Social	16	14
Escort and others	4	8

this information should be combined with the insight given by Watkins (1984) into the behaviour of the 'committed' cyclist. Both Watkins (1984) and the ECMT (1986b) emphasize the significance of cycling accidents as a deterrent to the cyclist making certain kinds of journeys, and to the cycle user incorporating the bicycle into a regular journey pattern. Table 7.11 shows the number of cyclists killed per million inhabitants in 1973–4 and 1980. In all countries, except Britain, there is a decline in cycling fatalities at the same time as bicycle use has increased.

The ECMT (1986b) reviews the measures available to reduce the number of cycling accidents and fatalities, and emphasizes the establishment of a traffic system in which conflicts between cyclists and motorized traffic are minimized. This is best done through a system of cycle paths.

Geographical variations in cycling accidents and in cycle ownership and use closely reflect national attitudes to cycling. In this respect, cycling has a significant advantage over the car as a policy area - it is clearly much more susceptible to changes in the level of use than the car and is, therefore, an effective policy instrument in urban transport planning.

In 1975, Holland constructed two 'demonstration projects' in the Hague and Tilburg to test various ideas about the design and impact of cycling facilities. The state subsidy to cycle facilities was increased from 50 to 80 per cent as a further encouragement to other local authorities to make special provision for cyclists.

Table 7.11 Number of cyclists killed per million inhabitants

Country	1973–4	1980
Netherlands	31	30
Belgium	28	25
Germany	23	19
Denmark	27	16
Sweden	17	14
Switzerland	13	12
France	13	11
Austria	23	10
Norway	10	7
UK	6	6
Ireland	4	3

Source: ECMT (1986).

Table 7.12 Cycle paths in German cities

Category (size of urban area)	Total length of cycle way (km)	Length of cycle way as percentage of overall street length
1,000,000+		
Hamburg	762.0	19.3
500,000–1,000,000		
Frankfurt am Main	211.6	19.6
Duisburg	298.7	25.1
Bremen	402.0	32.4
Hannover	296.8	26.8
300,000–500,000		
Mannheim	131.3	18.3
200,000–300,000		
Kiel	109.8	22.8
Oberhausen	96.7	21.0
150,000–200,000		
Mainz	75.5	27.5
120,000–150,000		
Oldenburg	92.8	20.2
100,000–120,000		
Erlangen	117.6	40.1
80,000–100,000		
Marl	66.9	22.0
Ratingen	45.2	22.5
Flensburg	48.5	25.5
Neumunster	52.1	22.1
60,000–80,000		
Bamberg	44.1	19.8
Delmenhorst	60.4	35.0
Russelsheim	35.8	22.3
Luneburg	60.4	32.9
Norderstedt	57.8	47.1
40,000–60,000		
Menden	90.0	35.5
Bergkamen	66.9	35.7
Waiblingen	68.3	56.9
Stade	37.0	19.4
Erftstadt	106.7	46.1
Elmshorn	24.4	22.6

Source: Hoppner (1979).

In German cities grants are available under the local government financing law for up to 85 per cent of the cost of constructing cycle tracks. For cycle tracks on roads, which are the responsibility of the *Land* (region) or *Kreis* (district), grants are available for up to 50 per cent of the cost. Table 7.12 shows the extent of cycle paths in German cities.

In the UK attitudes of government have been less encouraging. A governmental consultation document on cycling (Department of Transport 1981) was lukewarm and failed to provide the positive support for cycling which can be justified in terms of land use and congestion costs. However, the UK government will pay 50 per cent of the implementation costs of large-scale urban cycleways (Davies 1984).

The provision of cycling facilities encompasses a wide range of options, ranging from the purpose-built, segregated cycle track at one end of the spectrum to the provision of advisory cycle routes on existing highways at the other. Bicycle parking facilities may be provided in town centres and pedestrianized zones can be designed to make the task of the cyclist easier (e.g. Groningen in the Netherlands, and Darmstadt in West Germany). While a great deal can be done for the cyclist outside segregated cycle-way facilities, it is this topic which is most commonly associated with positive action for cycling.

Cycle tracks require careful design to fit in with the other aspects of transportation in urban areas; they should, for example, reflect the dominant flow lines and connect important sites, such as bus and railway stations, with town centres and with residential areas. They should be designed to free the cyclist from traffic hazards by utilizing features such as riverside paths, canal towpaths, abandoned railways and pedestrianized areas themselves (as in Darmstadt). Careful design can provide quick and safe routes through congested urban areas and give cycling a real advantage in competition with other modes. This will involve a careful reappraisal of existing traffic routeway and traffic management practices as much as the grafting of new cycle routes on to an unchanged traffic system.

Hudson (1982) describes some characteristics of cycle tracks besides main roads:

1. Junctions with major roads should be grade separated or have traffic signal control
2. At junctions with minor roads, the cycle track should continue across the mouth of the road and priority should be given to cyclists
3. If the cycle track rejoins the road, it should do so at an angle of less

than 15° to the road kerb and should be followed by a short section of cycle lane

4. At T-junction entering the road on the opposite side to the track, protected junctions should be provided for cyclists
5. Motor vehicles should be physically prevented from using tracks by raised kerbs and bollards
6. Cycle tracks should pass behind lay-bys and bus stops
7. Regular maintenance and sweeping schedules must be adopted and arrangements should be made for salting and clearing snow in winter
8. Clear signalling should be provided in positions easily read by cyclists.

Cycle planning requires this level of detail if it is to succeed, as well as a great deal of ingenuity to provide routes for cyclists through existing urban centres. In Middlesbrough, in north-east England, new cycle ways have been constructed leading towards the town centre from peripheral housing areas, but they experience considerable difficulties in leading the cyclist through the existing street network with the same degree of protection as that provided by the cycle way. In Lancaster, in north-west England, three different recreational cycleways on the track beds of abandoned railways converge on the city centre but then stop suddenly, leaving the cyclist in very dangerous traffic conditions on the edge of a one-way gyratory system.

Grimshaw (1982) advocates the use of 'soft routes' in the UK as a means of extending cycle facilities mainly for recreational purposes but with some urban components. His estimate of 14,700 km is arrived at from a study of disused railways, waterways, canals and rivers, public footpaths and bridleways, and forestry roads. The costs of such cycle ways or 'bike paths' are not great. The Ashton–Pill Project cost £4,174 per kilometre (1981 prices), and the Bitton–Bath railway section £1,605 per kilometre (Grimshaw 1982).

In a survey of county council cycle planning, in the UK (Davies 1984), it was shown that English county councils planned to spend in the region of £2 million on cycle facilities (0.2 per cent of their total capital bids). Of the thirty-nine English shire counties, eighteen (46 per cent) had not specified any sum for cycle facilities in their financial year 1984–5 bids for funds. This is clearly inadequate for the purpose of building new cycle ways and solving the problems of cyclists among heavy traffic in city centres.

The European-level response to the needs of cyclists has seen some recent signs of development. The ECMT (1986b) contains a long discussion on the safety of cyclists which, in passing, recognizes that cycling has important advantages in the areas of 'spatial use, energy saving and environmental protection'. Promoting the bicycle is encouraged and is linked to measures, such as cycle-way construction, which can guarantee the safety of cyclists. The document recommends several measures aimed primarily at preventing accidents:

1. Construction of cycle paths to segregate cyclists and motorized traffic
2. Improve technical standards of bicycles
3. Improve lighting and conspicuity of bicycles
4. Rider training
5. Safety campaigns and enforcement.

This theme is taken up by the European Parliament (1986), in its identification of risk factors for the cyclist. These include bicycle standards, visibility, behaviour of cyclists, and cycle tracks. Cycle tracks based on full segregation are preferred, but where this is not possible, clearly marked cycle ways and cycle routes in existing roads should be declared. The report also identifies the needs of cyclists in pedestrian areas and on one-way streets, and in terms of their use of public transport. Effective integration of the bicycle with public transport is seen as most important, including adequate facilities for carrying bicycles on trains and buses. In the UK the attitude of British Rail (BR) to carrying bicycles has been a set-back for recreational and commuter cycling, and the clear statement of a European preference here is indeed helpful. International public transport for bicycles is unsatisfactory, and the European Parliament would like to see improvements on trains and on aircraft, and in customs regulations controlling the passage of bicycles across national frontiers.

The use of the bicycle as a means of transport in urban areas, and as an important means of recreation, is a growth area in transport. Cyclists have particular needs which can be met at a cost which is a fraction of that associated with road construction. With careful design and planning, cycling facilities can be incorporated into urban planning, transport planning and conservation to bring about a better quality of life and an improved urban environment more in tune with Europe's historic cities and dense urban population. The potential is clearly there, and European-level institutions have grasped this point. It is possible that the

CTP could embrace bicycle transport and make a real contribution to the daily lives of Europe's transport consumers, bringing together land use planning, transport planning, and planning for a better environment, and thereby demonstrating that there is a genuine physical pay-off for European level intervention in planning matters.

8 Maritime, inland waterways, and air

Maritime, air, and inland waterway modes of transport are collected together in this chapter for convenience. There is a clear functional relationship between maritime policies, particularly those affecting ports, and inland waterways, and these two areas will first be dealt with. Air transport functions more as a self-contained transport system, though it does raise important issues of relevance to other modes, particularly surface transport in and around major international airports and their metropolitan partners. In all three modes there are important policy issues which link regulation and deregulation with the high costs of investment in expensive physical facilities, such as airports and seaports, and identify the need for integration. Seaports can only function effectively as part of an integrated land–sea transport system, where road, rail, and inland waterway links serve the port and its traffic and the port handling systems are up to the tasks of serving a shipping fleet and their specialized cargoes. Structural change in dominant trading routes, methods of organization in the shipping industry, and cargoes and cargo loading technology, all impose great strains on this industry and frequently necessitate some kind of state involvement. Air transport continues to fuel a demand for extra airport capacity which in some cases, as in south-eastern England, imposes a great strain on the environment, land-based systems, and planning procedures. Inland waterways can function only within a fairly sophisticated system of international regulation which has stood remarkably well the test of time.

Within the EEC there are marked differences in attitudes to these transport policy areas and accidents of geography will ensure that the UK, for example, has less interest in inland waterways than Germany and the Netherlands; Greece will be interested in maritime matters and those countries like the UK, France, and Germany with large 'flag-carrier' airlines will be most interested in air policy.

Of the three modes of transport discussed in this chapter, only inland

134

waterways fall within the original remit of the Common Transport Policy (CTP). Air and maritime have been added later, in response to the interests of new entrants and to changed perceptions of transport. The EEC (1985c) states that shipping policy is part of the CTP, as mentioned in article 3(e) of the Treaty, though much of the shipping policy derives its legal force from article 84(2).

Maritime transport

Historically, sea-borne trade and merchant fleets have been of great importance in the development of Europe's maritime nations. Only in recent years has European dominance of world sea-borne trade and shipping tonnage been challenged by the growth of 'open-registry' countries, and of Third World countries (see table 8.1).

Within the aggregate changes shown in table 8.1 there are significant variations in the type of shipping which is declining or increasing. One of the problems of EEC maritime policy is the seven-fold increase in the COMECON container fleet, which is in direct competition with the EEC container fleet. This rapid increase and other shifts in the balance of shipping tonnage between oil and bulk trades, general trades, and container trades cannot be inferred from the table.

Shipping is an important sector of the EEC economy. In 1982 member state fleets had net earnings of $9,100 million, spending $3,000 million on new tonnage, 44 per cent of which was ordered in Community yards.

Table 8.1 Merchant fleet of the world, 1970–83 (million gross registered tons and percentage of world total)

	World	EEC*		Open Registry†		COMECON‡		Others§	
			(%)		(%)		(%)		(%)
1970	211.9	64.9	30.6	40.2	19.0	13.0	6.1	17.3	8.2
1974	296.0	88.6	29.4	71.4	24.1	16.1	5.4	25.3	8.5
1978	386.6	108.3	28.0	105.3	27.2	21.5	5.6	47.8	12.4
1983	400.0	93.3	23.3	105.1	26.3	24.7	6.3	77.2	19.2

*EEC of the nine (i.e. excluding Greece, Spain, and Portugal).
†Liberia, Panama, Cyprus, Somalia, Bermuda, Bahamas.
‡Socialist countries of eastern Europe.
§All developing countries including South Korea, Hong Kong, and China.
Source: 'Progress towards a Common Transport Policy: Maritime Transport', EEC Bulletin Supplement 5/85.

Table 8.2 International goods traffic by merchant shipping, 1982 (1,000 tonnes) (loaded and unloaded)

	Total EUR10 (A)	Grand total (B)	A/B × 100
Germany	35,226	131,839	26.7
France*	–	–	–
Italy	26,476	238,073	11.1
Netherlands	75,560	317,361	23.8
Belgium	22,857	112,961	20.2
UK	104,243	241,692	43.1
Ireland	9,986	14,555	68.6
Denmark	13,186	36,639	35.9
Greece†	–	–	–
Total	287,534	1,093,120	26.3

*Incomplete data in Eurostat (1985), Transport and communication statistics.
†Greece is not included in the statistics.

The Community is the world's biggest trading area, accounting, in its trade with non-member states in 1982, for nearly 21 per cent of world imports and 20 per cent of world exports by value. Approximately 95 per cent of the tonnage of Community trade with non-member states and 30 per cent of extra-Community traffic is carried by sea.

Table 8.2 shows that merchant shipping in the EEC accounts for 287 × 10⁶ tonnes of international trading within the EEC itself, compared with 1,093 × 10⁶ tonnes traded globally by EEC countries (26 per cent). Within the EEC, different countries show varying degrees of dependence on Community shipping when compared with extra-Community trading. The UK and Ireland are responsible for the greatest percentage of tonnages loaded and unloaded which originate or are destined within the EEC.

Trading between partners in the EEC, in terms of tonnage lifted and unlifted, presents a complex picture (see table 8.3). The UK has consistently higher loadings in the outward direction than inward, with the Netherlands in a similar position; Germany is more mixed, with inward traffic from France, the Netherlands, Belgium, and the UK exceeding flows in the opposite direction. The UK accounts for the largest international flows by merchant shipping (within the EEC) followed by the Netherlands; in the case of global flows, the two countries exchange places in the rankings. It is clear, therefore, that while Europe is still a

Table 8.3 International goods traffic within EEC, by merchant shipping, 1982 (1,000 tonnes)

Legend: diagonal split of each cell — upper-left = unloading from, lower-right = loading to.

	Germany	France	Italy	Netherlands	Belgium	UK	Ireland	Denmark	Greece
Germany		650 / 389	142 / 265	7,014 / 2,499	1,102 / 617	11,393 / 3,041	166 / 250	3,591 / 3,904	43 / 159
France	397 / 675		2,987 / 6,887	4,219 / 1,837	1,219 / 740	20,475 / 10,843	422 / 932		1,454 / 1,291
Italy	296 / 176	6,779 / 3,265		2,748 / 1,629	939 / 652	3,246 / 1,507	19 / 63	66 / 109	1,196 / 2,986
Netherlands	2,459 / 7,040	1,919 / 4,260	1,111 / 2,691		300 / 1,277	27,097 / 20,783	489 / 1,587	600 / 2,578	793 / 570
Belgium	652 / 733	704 / 1,105	865 / 841	2,008 / 453		7,313 / 6,344	296 / 439	184 / 294	139 / 428
UK	7,308 / 14,155	5,651 / 12,162	2,995 / 3,347	13,560 / 20,025	4,938 / 5,518		1,332 / 5,339	1,514 / 5,600	468 / 329
Ireland	492 / 275	712 / 278	112 / 45	729 / 149	379 / 282	5,735 / 1,320		32 / 11	25 / 9

Source: Eurostat, Annual statistics, Transport, communication, tourism, 1970–83.

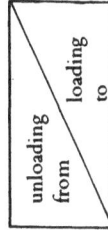

significant maritime trading bloc, the importance of this sector of the economy will vary widely from country to country, as will the degree of interest in maritime policy matters.

The EEC has been involved in the formulation of shipping policy only since the second half of the 1970s, with a submission to Council in 1976 of a communication on shipping matters and relationships with third countries. Prior to this, there has been an involvement with ports, particularly through the Port Working Group set up in 1972 (EEC 1977). Article 84 of the Treaty of Rome gives the Council power to decide on provisions for sea transport, and some general principles of the Treaty relating to monopoly and preferences may be interpreted as applying to shipping. Couper (1977) discusses the EEC's opposition to the United Nations Conference on Trade and Development (UNCTAD) Code of Conduct for liner conferences. Introduced in 1974 for the regulation of liner conferences, the Code was unacceptable to the EEC because of its discriminatory provisions which conflict with the Treaty of Rome. France, Germany, and Belgium, in supporting the Code, were in contravention of the Treaty. In 1979 an agreement was reached which permitted accession by its member states to the UN Code of Conduct (Regulation 954/79), and the Code entered into force in October 1983.

The EEC (1985c) lists the basic principles which should 'guide further Community action in shipping':

1. The predominant issues affecting shipping are those concerning trade with third countries; the international competitive position of Community shipping is of significance
2. Wide international agreement is to be preferred to unilateral Community initiative; this is necessary to protect EEC interests outside the Community
3. Common policy action on shipping matters should be matched by equality of treatment of Community shipowners
4. The Community should continue to pursue a non-protectionist shipping policy based on the principle of free and fair competition in world shipping
5. The Community should support international efforts to maintain and improve maritime safety
6. The Community should seek to improve the commercial competitiveness of Community shipping and thus contribute to such general objectives of the Treaty as economic development and the improvement of employment opportunities in the shipping industry.

The EEC has been particularly active in the area of 'cargo reservation' – i.e. attempts by some national governments and shipowners to reserve, for their own carriers, a proportion of the available trade. These tendencies are resisted by the Community, and it is an area where Community action is likely to be more beneficial than actions taken by individual states. Community action can ensure that cargoes are not simply diverted from a port in one Community country to a port in another. The Community is particularly concerned about unfair practices as they emerge in the activities of state trading countries, such as the USSR. These unfair practices – underquoting and creaming off high-paying cargo – are causing damage to Community carriers. The EEC is also active in maritime safety and the prevention of pollution, 'port state control' which monitors ships using community ports, and the provision of navigational assistance in European waters.

The Community has recognized the importance of seaports in reports which go back to 1961, but particularly in the Report of the Port Working Group (EEC 1977) and the Carossino Report on the role of ports in the Common Transport Policy (EEC 1983a). Ports are particularly important points of convergence in the transport network of all states and within Europe have an additional importance, given the amount of freight which travels by inland waterway and ends up in this port system. Ports are also important geographical areas, with a distinctive service-based economy and a heightened vulnerability to structural changes in shipping. As such, their consideration within the EEC needs special care, to bring about a satisfactory match of regional development, social competition, and maritime policies. The EEC discouragement of policies which infringe competition rules, such as those which subsidize port operations and investments, is likely to produce severe social and employment problems in areas with little alternative employment. The EEC activity in other areas, especially agriculture, also has the effect of depressing shipping and shows a markedly different attitude to shipping than the protectionist policies displayed elsewhere. The EEC (1985c) outlines in detail draft council regulations to come into force in July 1986, which are aimed at safeguarding free access to cargoes in ocean trades. Also tackled are unfair pricing practices which work against the interests of Community shipping. Another area of EEC intervention in shipping matters is through its competition policy. The principles governing competition in the Community are laid down in articles 85 and 86, which prohibit agreements preventing or distracting competition. Certainly this affects maritime and air transport, and both have lagged behind in the application of Community competition policy. In shipping both liner conferences and loyalty agreements are in breach of EEC competition rules. In 1981 the Commission proposed a draft regulation which would have served the purpose of bringing shipping into line with competition policy (EEC 1981).

This regulation would grant most liner conferences and loyalty agreements exemption from the ban on restrictive agreement imposed by article 85, but this could be subject to conditions. It would also provide for the means of enforcing the regulation through appropriate courts. The details are discussed by the House of Lords Select Committee on the European Community (House of Lords 1983a).

The nature of EEC intervention in shipping is designed to protect Community shipowners, shipping's share of the market, and the economic value of this sector of the economy. Given this, and the traditional isolation of the world of maritime policy from other aspects of transport, this area of EEC policy is relatively coherent and relatively focused, in a way that most of the policies on land-based systems are not. There is a communality of interest here, which is likely to bring larger pay-offs than would be available to a member state acting in isolation. The EEC, in spite of the decline of its fleet, is a large trading bloc with a large merchant marine. Its ports are among the best developed in the world and it is in a position to influence global international organizations, as well as non-EEC countries, whose activities may be contrary to Community interests. In spite of conflicts and internal contradictions, the maritime transport policy of the EEC stands in relatively good order and shows the benefits of a Community response.

Inland waterways

The inland waterways of Europe represent an important and long-standing mode of transport, of more importance to the original six member states than to subsequent additions. This network can be seen in figure 8.1.

The extent of the inland waterway network is such that it connects the Mediterranean with the North Sea, and Channel ports with eastern Europe (with the opening of the Rhine–Main–Danube Canal in 1992). This is a valuable asset for European economic development and European freight flows, and one capable of a far more important role in the overall transport system of an enlarged Community.

Table 8.4 shows the distribution of navigable waterways among

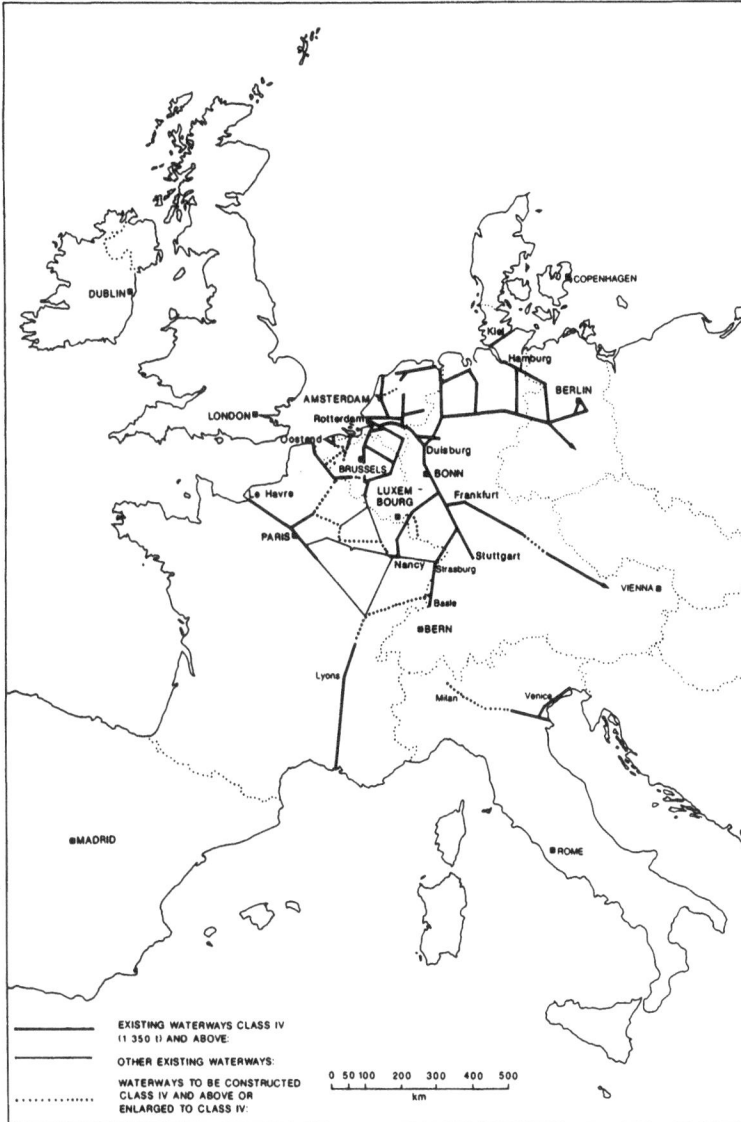

Figure 8.1 Map of Europe's Inland Waterways

Table 8.4 Navigable waterways in Europe, 1983

	Length of canals (km)	Length of rivers (km)	t km × 10⁹ performed*
Germany	1,000	2,097†	49.09
France	918	1,589	9.45
Netherlands	169	240	32.33
Belgium	216	499	4.97
Totals	2,303	4,425	95.84

*Includes canals and rivers.
†The Rhine accounts for 622 km.
Sources: Eurostat (1985); ECMT (1985), p. 75.

member states, thereby identifying those with interests in this mode of transport. Germany has the largest stretches both of canal and river and also the largest total of tonne-kilometres handled. In 1982, Germany's navigable waterways were responsible for 49.09×10^9 tonne-kilometres. The Rhine provided the majority of this total, and if the Rhine is considered as an entity from Basle to the North Sea, it is responsible for 54.4×10^9 tonne-kilometres or 57 per cent of all the tonne-kilometres on Europe's navigable inland waterways.

Other European countries have a much lower dependence on inland waterways. In the case of Britain, with a sizeable length of navigable waterway (4,830 km), of which 541 km is designated 'commercial waterways', little freight is carried by European standards: 2.4 billion tonne-kilometres in comparison with Germany's 52 billion tonne-kilometres (1984). This does not, of course, preclude the development of inland waterways towards a greater role in freight transport.

Van der Bos (1983) has carried out a review of the inland waterway transport market in western Europe, where he shows that the conditions under which operators function vary markedly from country to country. There is a problem of over-capacity and, according to the same author, there is little prospect of an increase in demand for inland waterway transport. With a decline from 80,000 employees in 1965, to 40,000 in 1982, the industry has already adapted to changed circumstances and is aware of the potential for new development in more efficient boats, improved waterways, and an enhanced role as the real social costs of road freighting became more apparent. Either alone or in various combinations with other modes of transport, inland waterways can offer an energy-efficient and relatively pollution-free mode of transport.

The objectives of the EEC with respect to inland waterways have been set out in a European Parliament report (EEC 1979):

1. Reduce the economic and structural disequilibrium of supply and demand for tonnage
2. Reduce the size of the fleet and adapt it better to the requirements of users through qualitative improvement
3. Harmonize national and international measures
4. Prevent distortion of competition by state-trading countries
5. Improve working conditions
6. Establish free market economy traditions in the transport market.

These objectives convey the clear imprint of the CTP, especially as regards the free market impetus and the harmonization emphasis. Inland waterways present an administrative challenge to the EEC, given clear national policies on capacity and scrapping of surplus carrying capacity. These schemes involve state finance and are little different in principle to industrial restructuring schemes in textiles, shipbuilding, iron and steel, etc., which attract a great deal of Community attention.

Chini (1980) presents a detailed analysis of the different national regimes governing the control of inland waterways, and the competitive

Table 8.5 Benefits of inland waterways

Direct:
1. Waterways handle about one-third of the total volume of traffic in Europe at comparatively low infrastructure cost
2. There are transport cost savings for the consignor in both domestic and international traffic
3. Use of waterway craft as (inexpensive) storage capacity

Indirect:
1. Lower consumption of primary energy than road or rail
2. Less labour cost to the state for traffic control
3. Lower accident rate
4. Less pollution
5. Less risk of loss through acts of war or terrorism
6. National supply assured by national transport
7. Work for shipyards and their suppliers

---

portion of inland waterways with respect to other modes and pricing conventions. The same study compiles a list of direct and indirect benefits of inland waterways (table 8.5).

The future development of inland waterways in Europe is likely to be heavily influenced by trends in three different areas:

1. The opening and development of new traffic on the Europa waterway (Rhine–Main–Danube)
2. The increased importance of the sea–inland waterway interface
3. Infrastructure investment through removal of bottlenecks and upgrading of existing links.

The first of these trends raises the question of state-supported operations, and the EEC is concerned about the influx of COMECON shipping on Community waterways, operating at artificially low prices and exacerbating an existing problem of over-capacity and structural change. The Community response to these changes has not yet been formulated, but much will depend on Germany's attitude to bilateral agreements with other countries over the use of the River Main as the link in the system. On the positive side, there is a large potential for trade with eastern Europe and the Soviet Union, and the Europa waterway could play a major role in the development of this traffic and in associated land-based industrial developments. The second trend will be dealt with in chapter 9, where the topic of combined transport is examined in detail.

The third trend is one which for many years has exercised the CTP and is equally relevant to all modes of transport. The EEC Directive 83/643/CEE, the 'Frontier Facilitation Directive', has been agreed, and member states agreed to implement it by the end of 1984. Similarly, agreement in principle has been reached in the Single Administrative document, which would bring together the seventy-plus different forms currently in existence for processing imports, exports and transit traffic. While of some importance to lorries and railways, such measures are also of economic benefit to inland waterways. The Commission of the European Communities has already made thirty-two separate proposals to harmonize the veterinary, public health and plant health regulations which will simplify border checks.

There are particular checks at frontiers which affect inland waterway vessels, cause delays, and impose extra cost. These include checking and levying duty on fuel, licensing of vessels for the carriage of dangerous goods, and the collection of VAT on transport services. Optional checks can add the collection of navigation and canal dues, checking compliance with police regulations on manning and checking navigability certificates. As Rhine navigation has been liberalized by the Mannheim Convention (1848) and other agreements, frontier barriers are not as onerous as they might be, but crossings between Belgium and France, or Belgium and the Netherlands, do present problems in need of some alleviation.

Inland waterways represent an especially successful transport mode both in terms of low social cost and high degree of liberalized and integrated transport. Its structural problems are not great by comparison with the potential for developing new markets, particularly in eastern Europe and over the link to Berlin, through East Germany. In combination with a vigorous approach to marketing, the advantages of waterways jointly with their seaport terminals, and a modest programme of investment in infrastructure improvement, inland waterways could carry a much bigger share of Europe's freight, at a much reduced environmental and energy cost.

Air transport

Air transport has presented serious problems for EEC transport policy makers and was excluded from the provision of the Treaty of Rome, in spite of its designation as a 'priority area' in the Spaak Report which laid down the framework of the Treaty of Rome.

Civil aviation in Europe is dominated by large national carriers which are predominantly state-owned and seen as representatives of that country abroad, and important assets both militarily and in terms of a foreign policy support role. It is to be expected therefore, that the larger member states of the EEC have taken a view of their airlines which is very much outside the view they would take of other sectors of the economy, and of competition policy generally. Within the EEC there can be little doubt that current patterns of regulation and organization in the airline industry are at variance with EEC competition policy (articles 85 and 86) and general statements on the free movement of goods, persons, and capital within the Community.

European civil aviation is regulated by a set of bilateral treaties known as Air Service Agreements (ASAs), entered into by governments and controlling market entry, tariffs, routes, and capacity. The ASA produce a highly regulated industry (McGowan and Trengove 1986) with two dominant features: (a) Entry on to individual routes and, in some cases, all

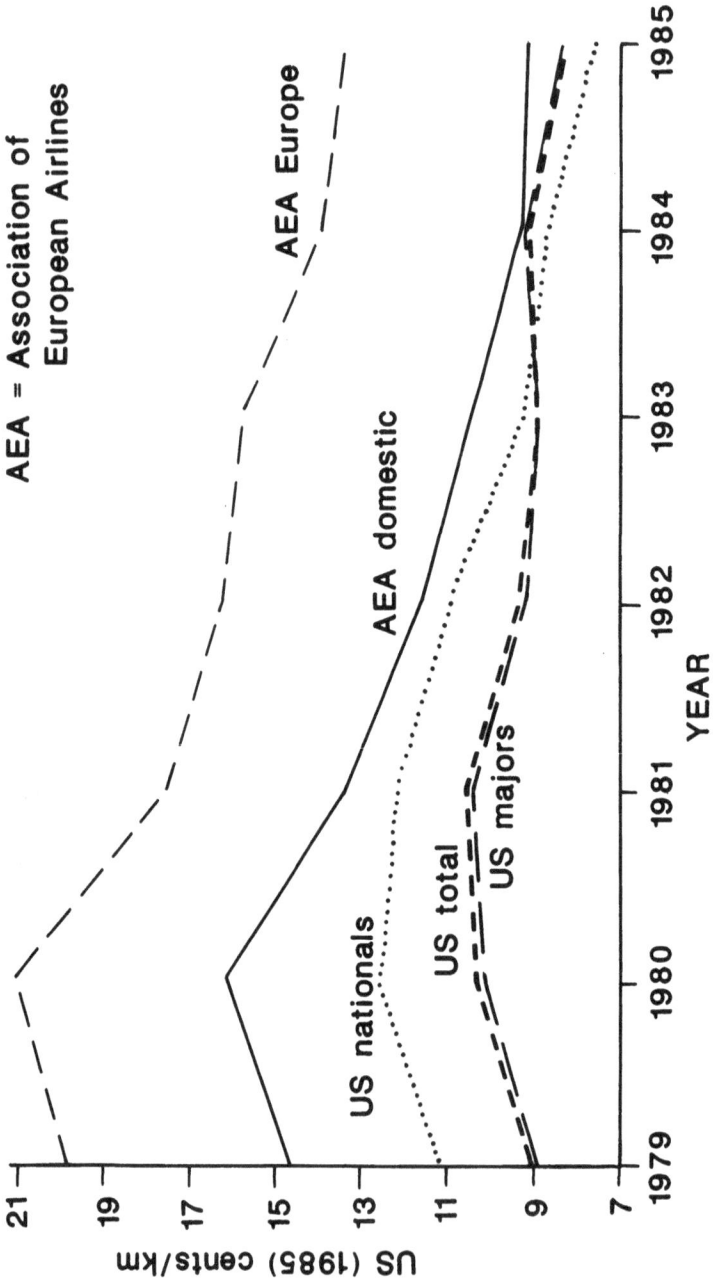

Figure 8.2 European and US income by length of journey (AEA is the Association of European Airlines)

routes between the two countries is usually restricted to a single 'designated' airline from each country, and such markets can be termed
statutory duopolies; and (b) The regime encourages arrangements
between the duopolists aimed at co-ordinating their activities; the result
is the elimination of competition from most scheduled airline markets in
Europe.

The advantages of regulation, such as they can be identified, lie
mainly in the area of protecting established national carriers and thus
supporting some or all of the intangible benefits associated with their
market dominance. As in all regulatory systems, it could be said that a
degree of market stability and high safety standards also result from
regulation. The evidence of US deregulation, coupled with the continuing existence of international agencies with an interest in safety,
suggest that regulation cannot really benefit the consumer to the extent
that its supporters suggest. Regulation does permit a substantially higher
level of fare than would be possible in a deregulated system. McGowan
and Trengove (1986) suggest that European fares are 45–74 per cent
above those in the USA, and this differential cannot be explained by
reference to factors other than the absence of competition in Europe.
Indeed, US fares could be expected to be above European fares as a result
of labour cost differentials.

Figures 8.2 and 8.3 show the USA–European fare differentials.

The disadvantages of regulation relate mainly to the fares issue and are
succinctly put in the UK by the Consumer Association in its evidence to
the House of Lords Select Committee on the European Community
(House of Lords 1985a):

> Air fares are excessively high and the range of services offered per
> route are restricted. National airlines have been protected from
> competition at the expense of passengers who have been denied the
> choice of services at reasonable prices that would otherwise be
> possible.

The attitudes of European Airlines are expressed through the Association of European Airlines. The Association opposes US-style deregulation, while emphasizing that the present system could be more flexible:

> On a number of points AEA member airlines share the conclusions of
> the Commission, notably the desire for greater regulatory flexibility,
> while maintaining the benefit of inter-airline co-operation, including
> multi-lateral tariff co-ordination and the rejection of US style

deregulation as a regime suitable for application in Europe.
(House of Lords 1985a)

EEC progress with aviation policy has been painfully slow, reflecting
no doubt the entrenched position of large member states with veto
powers. It was in 1974 that aviation and shipping policy were confirmed
as subject to the Treaty of Rome's general rules, and in 1975 that the
Commission proposed a 'Council Decision concerning the operation of a
common policy in the civil aircraft and aviation sector'. Nothing came of
the 1975 initiative. It was not until 1979 that a comprehensive examination of aviation policy was produced (Memorandum 1).

Memorandum 1 ruled out full freedom of entry. Airlines seeking
access to a new route would have to satisfy certain conditions (e.g. on
profitability). The Commission called for the development of traffic
between regions, a review of the structure of bilaterals, and an examination of state subsidies in air transport. The Memorandum did not get to
grips with competition policy and its non-implementation in the
aviation sector. Airlines in Europe were still blatantly out of line with
EEC competition policy and the European Parliament picked this up in
its criticism of Memorandum 1. McGowan and Trengove (1986)
concluded that the Commission had very little to show for its efforts as a
result of Memorandum 1.

The Commission's next action was more inspired and took the form
of a proposal to 'free' a particular sub-market of air transport from
regulation. This was in fact selective geographical deregulation and
concerned inter-regional transport on non-trunk routes. The aim was to
allow free entry on to routes between category 1 airports and major
provincial airports and the airports with international freight capabilities.
Aircraft with up to 130 seats would be permissible and there would be a
minimum stage length of 200 km. New entrants on such routes would
have considerable flexibility in pricing their services.

This proposal was inspired in so far as it is one of the few which gets to
the central problem of transport services for many groups of users - i.e.
their spatial concentration and relative inaccessibility. A regulated
aviation system had not produced a variety of services outside the most
important national airports, even at the domestic level. This concentration at Heathrow, Gatwick, Paris, Frankfurt, and elsewhere, at the level
of the largest cities, has had a potent effect on the development (or lack of
development at regional airports, on the length of journeys to airports for
passengers, and on congestion and discomfort at certain airports, notably

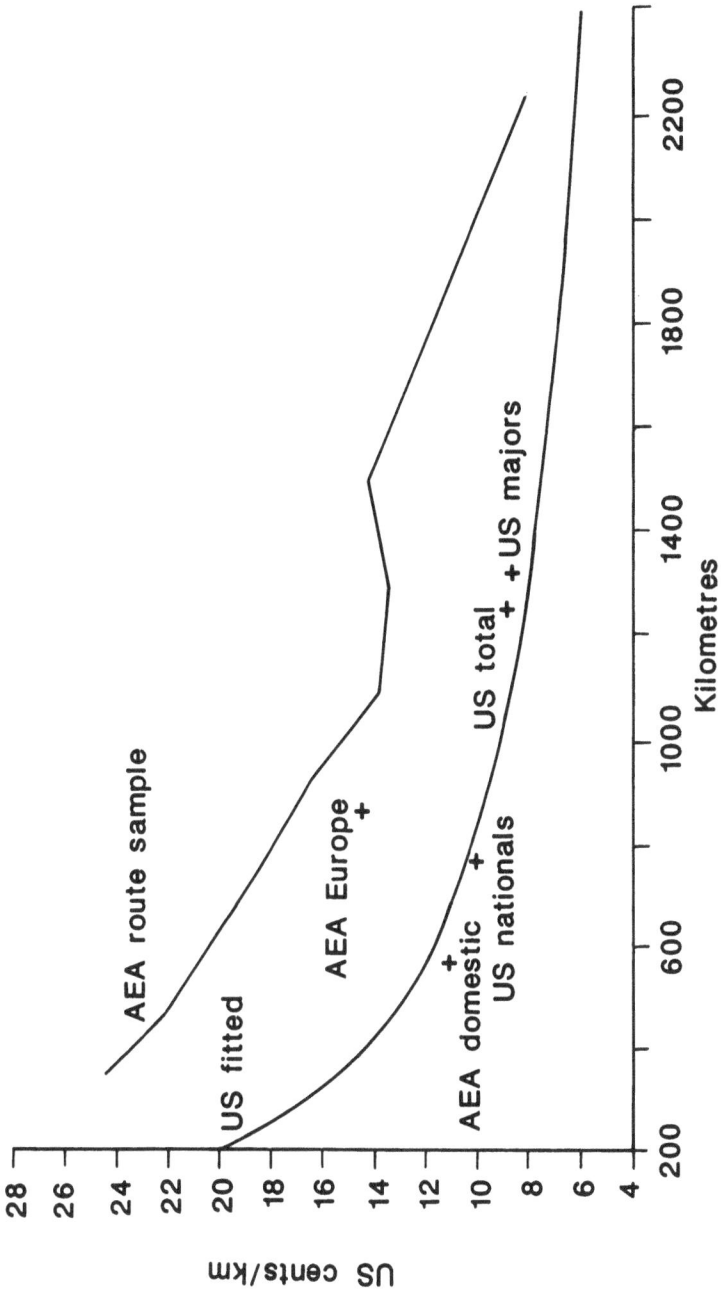

Figure 8.3 European and US income in constant 1985 prices

Heathrow. Around airports such concentration also leads to excessive environmental nuisance. The Commission's attempts to produce a selective geographical deregulation held the promise of a larger number of air services, connecting a larger number of destinations with attendant benefits for users, travel costs, and environmental impact on selected airports.

This point can be supported by the evidence given by Manchester International Airport Authority to the House of Lords Select Committee (House of Lords 1985a). Manchester can point to the fourfold increase in use of services when a direct route from Manchester to Australia was provided, and to the growth of 'interlinking' via overseas gateways as UK passengers transfer to further destinations outside the UK (e.g. Amsterdam), as evidence for latent/unsatisfied demand at regional airports. The effect of regulation is to disadvantage services away from Heathrow Airport. Pakistan International, Malaysian Airlines System, Singapore Airlines, and Air Laker have all either been refused permission to fly out of Manchester or been put off in informal discussions. Manchester concluded its evidence to the Select Committee: 'The initiatives of the European Community to relax regulatory control are to be welcomed in general, as by relaxing the requirements for capacity to be equalised, the way would be open for services to be operated from alternative gateways by airlines of either country.'

The EEC's regulation on inter-regional transport was supported by the European Parliament, the Economic and Social Committee, the European Regional Airlines Organization, and the independent air companies and consumer groups. The major airlines and the aviation unions opposed it. In Council discussions the proposal was considerably watered down. The UK and the Netherlands were supportive but opposition elsewhere led to dilution, and its re-emergence in Memorandum 2 of March 1984. Memorandum 2 sets down the Commission's aims which, until such time as a common air transport market (akin to the US situation) can be brought into being, seek to relax the existing system to give: 'a wider variety of choice for the consumer, lower costs and therefore lower prices, more scope and more profits for the efficient and innovative airline and a growth stimulus to the air transport industry.'

The Commission's proposals refer solely to inter-Community air transport, and can be summarized as follows:

1. Removal from bilateral agreements of any obligations placed on

airlines by governments to participate in arrangements which are inconsistent with the competition rules of the Treaty

2. Relaxation of any rigid adherence to a 50–50 split in capacity shares
3. Flexibility with tariffs via 'zones of flexibility', governments and airlines would set reference fares and zones on the basis of their traditional 'double approval' procedure, but within the zones, tariffs would become subject to either country of origin approval or 'double disapproval'
4. Exemption of various fare setting, capacity sharing and revenue-pooling activities from the competition rules for a fixed period of seven years, subject to certain conditions.
5. Greater transparency of state aids to air transport should be brought about, with a view to applying the Community's rules on subsidy.

Taken together, the proposals mean that if governments will agree to reduce controls on capacity and tariffs, the Commission will exempt its airlines from the full rigour of the competition rules of the Treaty. In effect, they could be exempted under the terms of article 85(3) and this portion would be reviewed at the end of a seven-year period. The Memorandum received a very mixed reaction from EEC organizations, industry, and consumer groups and subsequently has been influenced by important litigation.

On 30 April 1986, the Court of Justice delivered a judgment in response to a request for a preliminary ruling from a French court, confirming that the competition rules of the Treaty of Rome apply to air transport. According to McGowan and Trengove (1986:127) there is still a difficulty: 'The difficulty ... lies in the absence of specific rules for the implementation of Articles 85 and 86'. The result of this difficulty is that the price fixing activities of the European airlines have been preserved intact. Nevertheless, the Commission sent a letter to ten airlines giving them two months (from July 1986) to provide evidence that they do not participate in the cartel practices and sharing arrangements of the main European carriers. As a result of the judgment, the Commission foresees an agreement on arrangements for tariffs and fares capacity, and the application of Community competition rules under articles 85–86 of the Treaty of Rome.

Individual countries in the meantime are making progress towards a more liberalized air transport system through bilateral agreements. This includes agreements between the UK and the Netherlands, Belgium, Luxembourg, France, and Germany, and involves the use of regional airports, smaller aircraft and a more diversified pattern of routes based on the regional airports (e.g. Birmingham–Stuttgart, using Jetstream 31 aircraft, or Southend–Le Touquet with a Britten Norman 2A Islander). These services bring a significantly greater degree of choice to the air traveller and also distribute the benefits and disbenefits of air transport over a much wider area.

The UK has been the prime mover in bilateral liberalization. In June 1984 it negotiated an agreement with the Netherlands, whereby any carrier designated by either government would be allowed to fly any route between the two countries. The carriers themselves would decide on frequency and capacity of services. Similar agreements have been reached with Luxembourg, West Germany, Italy, and more recently, Spain, where the existence of charter operations has led to the conversion of unscheduled services into scheduled services. The new bilateral agreements appear to have been successful. On the UK–Netherlands route as a whole, capacity growth in the last year of the pre-bilateral period (June 1983 to June 1984) was about 5.9 per cent. In the first year of the new regime capacity increased by 24 per cent. The effect on fares is less clear, though it could be argued that the full effect on fares could not really be felt until full multilateral liberalization appears, as opposed to a number of bilateral arrangements, each with slightly different terms and conditions.

9 Combined transport

It is quite usual in treatments of freight transport, to regard major modes (road, rail, and inland waterway) as independent and competing with each other for a share of available freight. The purpose of this chapter is to illustrate the development of multi-modal or combined transport, where different modes are used for different legs of one consignment, within a scheme of general supervision which organizes the shipment from origin to destination. Under this arrangement different modes are complementary, not competitive.

Frybourg (1984) defines combined transport as:

The transport of integral loads for hire/reward with haulage units using at least two land transport modes ... Combined transport services relate to three types of traffic with quite different characteristics:
- The land transport of ISO maritime containers.
- The transport of different types of unit load for inland customers.
- Routeing by rail of vehicle elements belonging to road hauliers: Swap bodies, semi-trailers and, more rarely, complete vehicles with tractor units.

The advantages of combined transport are:

1. The use of different modes for carefully selected legs of one shipment can produce lower costs of freight movement and other economic advantages through the 'pipeline effect' - using the transport system as a storage system; and
2. The use of modes in combination can intensify modal utilization and so produce better returns on infrastructure investment, savings on investment avoided elsewhere, and large savings in land-take and other wider costs which result from duplication of modes in major transport corridors.

3. The avoidance of environmental disbenefits which accrue from
   excessive development of one mode, e.g. road through Alpine passes
   and concentration of traffic around major urban motorways as in the
   Frankfurt area; combined transport can offer attractive possibilities
   for bringing about environmental improvements at no detriment to
   transport efficiency, and with the possibility of an improvement to
   existing levels of utilization.

Seidelmann (1984) shows how combined transport can provide cost
savings for suppliers and solve transport problems associated with
crossing Switzerland and using a fewer number of drivers more effectively. Sciarrone and Carrara (1984) produce similar results (see figure
9.1) and show the cost advantages of combined transport over different
distances. These authors claim that the two sets of costs equal each other
at around 450–500 km, but give combined transport a clear advantage
over greater distances.

The key to releasing the energy locked in combined transport is the
road haulier and his/her perceptions of efficiency, costs, and control. The

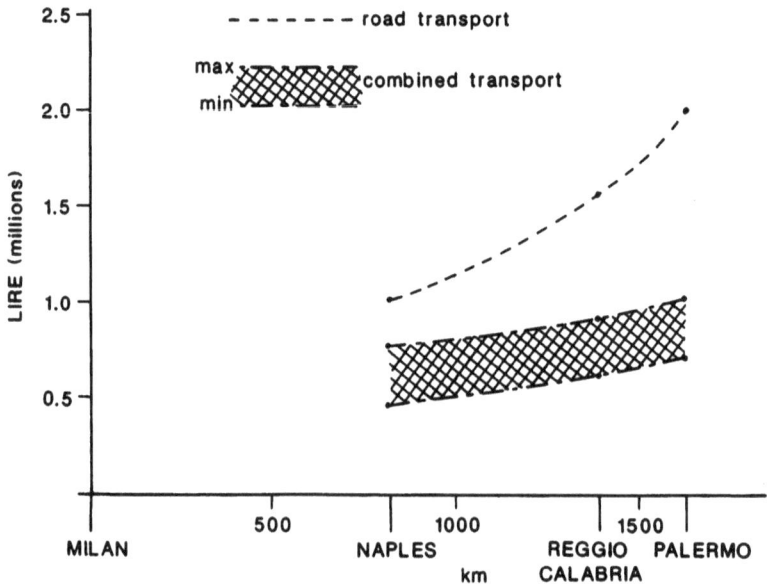

Figure 9.1 Cost advantages of combined transport. Costs shown are for
shipping one full load by combined and road transport over set distances in Italy

EEC can influence tariffs and improve efficiency through technical harmonization and infrastructure investment, but it cannot persuade road hauliers to sacrifice their control of the shipment from origin to destination if they fear that the alternative is not reliable. If they are protecting quotas and privileges, they may be even more reluctant to make the leap into what is a new organizational form, grafted on to old technologies.

There may be some unresolved problems lurking in the combined transport undergrowth. If combined transport is very successful, it could conceivably win custom from the railways themselves and impose extra road movements near origins and destinations. In this sense, the extension of current combined transport practices to include own-account operators as recommended by the ECMT in 1986 (CM(86)4) may produce more lorries in conurbations and port areas with consequent losses to traditional rail haulage and revenue (and environmental damage).

Combined transport is well organized to reap enormous benefits from the Channel Tunnel. The experience already gained in Europe, on 'piggyback' transport (the carriage of road vehicles on trains) and rail journeys of 1,000–1,500 km, will be ideally suited for journeys from UK cities to major centres in Germany, Austria, Switzerland, and Italy. The fine balance of economic advantages, both spatially and sectorally, which will arise from the Tunnel is still unclear, but 'combined transport' (the use of more than one mode of transport for a single shipment involving the transfer of vehicles and/or container from one mode to another in an organized and pre-determined manner) will provide a lot of traffic for the Tunnel and potentially generate a great deal of economic and environmental benefit if the final arrangements for Tunnel traffic include the possibility of several 1,000–1,500-km journeys, with the Tunnel providing the most important gains in delivery times and costs.

Under a Council Directive of July 1982 the freedom of road transport from all licensing and quota procedures has been extended to combined transport, between member states, or waterway/road containers. Exemption from licensing and quotas also applies to road transport within a radius of up to 50 km from the inland waterways point of shipment or unloading.

The EEC is working to reduce administrative formalities in combined transport to a minimum and to negotiate with third countries to free combined transport from all licensing and quota procedures. This will include Scandinavian countries, Switzerland, Austria, and Yugoslavia,

Table 9.1 'Piggyback' transport, 1982

Country	National				International			
	Vehicles	Semi-trailers	Swap bodies	Total	Vehicles	Semi-trailers	Swap bodies	Total
Austria	+	+	+	8,675	+	+	+	25,168
Belgium	–	+	+	206	–	+	+	24,285
Finland	–	–	–	–	–	–	–	–
France	–	57,347	55,938	113,285	–	37,149	33,475	70,624
Germany	28,000	40,900	212,400	281,300	13,500	49,100	35,100	97,700
Italy	+	+	–	14,347+	+	+	+	100,957+
Netherlands	–	–	–	–	24	10,141	470	10,635
Norway	–	–	–	–	–	+	+	2,250
Spain	–	–	2,000	2,000	–	–	17,476	17,476
Sweden	–	7,100	11,000	18,100	–	2,100	1,000	3,100
Switzerland	3,840	+	+	8,228	+	+	+	50,619

Notes: Units transported.
1 semi-trailer
2 swap-body 7 m
1 swap-body 12 m
1 tractor and trailer unit.

Source: ECMT (1985).
– no figures available.
+ only total figures available.

and will ease the problem of transit in these countries as well as extend the range of operation of combined transport services.

The EEC is developing a supportive environment for combined transport in other ways, including:

1. Remission of road tax where vehicles undertake journeys by rail (Directive 82/603 EEC)
2. Infrastructure assistance to rail links with combined transport potential
3. Encouragement to railway companies and 'piggyback' operators to associate themselves with Interunit, the company set up to facilitate international 'piggyback' transport
4. Technical harmonization, e.g. definition of swap-bodies and their minimum technical specifications
5. Easing customs formalities, e.g. through Regulation (EEC) 902/80 which simplifies procedures applicable to the carriage of goods in containers
6. Liberalized road haulage tariffs and use of 'inclusive tariffs' for through journeys (Regulations 3568/83 EEC and 84/646 EEC).

Combined transport is a growth area in EEC transport. Long distances, the advantage of rail for trunk hauling, the ability to knit together water, rail, and road, together with cost advantages, conspire to produce an area of high growth potential. When this is linked to problems of transit countries, environmental and energy concerns, and the need for acceptable social conditions for drivers of road haulage vehicles, the advantages are multiplied. The obstacles that do exist are organizational, institutional, and infrastructural and all are capable of resolution within a framework of international co-operation. The CTP can provide such a framework and can provide a means of implementing decision making with the other organizations active in this area such as the ECMT.

National 'piggyback' transport figures show Germany to be the major user, with France at about half the German figure and other countries further behind by a huge margin. In international 'piggyback' transport Italy is in the lead, closely followed by Germany, France, and a number of other countries, though the gap between heavy users and light users is much less than in the case of national transport (table 9.1).

The area of combined transport has been a fertile one for international agencies operating in Europe. The European Conference of Ministers of Transport has been active in this area, as has the Inland Transport

Committee within the United Nations Economic Commission for Europe. The EEC has also been active.

In 1975 the Community issued a first, experimental directive with a view to freeing road transport from all quotas and licensing in connection with 'piggyback' traffic, i.e. transport on terminal delivery journeys to or from rail loading or unloading stations. A second Directive, in 1978, extended the applicability of the first to include containers. In 1981 this second Directive was made permanent.

There is very little statistical information on combined transport which would allow an accurate appraisal of its importance in Europe today. Statistics on freight transport still reflect the conventional view of independent and competitive modes of transport. One insight into combined transport can be had through statistics on the use of containers. Table 9.2 shows the number of containers handled in seaports and transported by rail. What it does not do is to detail the movement of containers by road and rail – or either mode in combination with inland waterways and seaports. It does, nevertheless, illustrate the quantitative significance of container transport and, among other things, its significance to the German and French railways. In terms of seaport activity, the Netherlands and the UK are at the top of the league with Germany not too far behind.

An example of the success of combined transport is the growth of Intercontainer, a federation of twenty-three European railway companies who combine forces to market their container services throughout Europe, and include in their operations maritime containers as well as land containers and swap-body traffic. In 1985, Intercontainer carried 904,803 TEUs (twenty foot equivalent units), including more than 320,000 TEUs across the Austrian, Swiss, and French Alpine passes. European cargo haulage is concentrated upon Marseilles-Fos, with large-scale rail haulage over considerable distances. The other main centre of the distribution system is Rotterdam.

Container transport is important on inland waterways, though this mode adopted containers much later than sea transport. Combined transport on inland waterways takes place mainly on the Rhine (table 9.3) between Dutch and German ports, Strasbourg, and Basle, and between the ports of Rotterdam and Antwerp and those on the Danube.

The transport of road vehicles, in whole or in part, on railway services is clearly very important to combined transport. It combines the advantages of both modes and is flexible enough to adapt to different circumstances and market shifts. Collectively known as 'piggyback' transport,

Table 9.2 Container transport, 1975–83

	Handled in seaports TEU × 1,000					Transported by rail TEU × 1,000				
	1975	1980	1981	1982	1983‡	1975	1980	1981	1982	1983‡
Austria*	393					28	95	110	114	132
Belgium		720	822	804				291	261	
Finland		125	135	140	151	17	37	37	40	44
France†		1,057	1,088	1,026		412	790	813	762	749
Germany	739	1,541	1,779	1,750	1,805	574	1,021	1,065	997	989
Greece*						2.1	1.6	4.3	3.3	
Ireland†	232	193	224	228		113	151	152	161	159
Italy						215	542	528	495	
Netherlands	1,138	2,011	2,191	2,275	2,370	135	269	264	245	245
Norway	20	64	77	80		18	34	34	32	
Portugal	91	140	148	157	169	3	19	20	12	19
Spain*	550	1,157	1,265	1,320	1,517‡	80	217	220	236	245
Sweden	174	225	238			230	275	290	270	
Switzerland	–	–	–	–		70	225	225	219	
Turkey		100	156	190						
UK	1,800	2,400	2,400	2,500		381	551	522	495‡	

*Numbers, not TEU.
†Numbers in seaports.
‡Estimated.
Source: ECMT (1985).

Table 9.3 Container transport on inland waterways, 1975–83 (TEU)

	1975	1980	1981	1982	1983	Remarks
Austria (Danube)	–	–	–	7,906		
France	84	4,172	5,233	6,600	5,176†	Only Rhine traffic to and from Strasbourg
Germany*		96,000	121,000	147,000		
Netherlands:						
Rhine†	10,000	90,000	100,000	100,000		
to and from Antwerp†			30,000	25,000		
Switzerland		822	680	931		

*Loaded containers only; estimated.
†Estimated.
Source: ECMT (1985).

this form of operation can function in at least three different ways: swap bodies, semi-trailer, and road trains. The road train, or in Germany *rollende Landestrasse*, is particularly effective over longer routes. According to ECMT (1985), the ten countries in Europe participating in 'piggyback' transport, achieved a total of 463,138 international 'piggyback' consignments.

10 The Channel Tunnel

The Channel Tunnel is probably the single most important development in European transport this century. Its significance is as symbolic as it is practical, but in a growing European Community seeking to become more unified and economically successful, symbol and reality are not easily disentangled.

The Channel Tunnel's contribution to European transport and the wider objectives which transport must serve lies in six main areas:

1. The linking of Britain's transport system with mainland Europe and eradication of a major bottleneck for both road and rail transport
2. The economic development possibilities which arise from such a stimulus, always bearing in mind the job losses which will result from technological change and the geographic restructuring of economic activity which might work in favour of northern France and to the disadvantage of Kent
3. The environmental consequences of a Tunnel, which if properly secured could bring about a transfer from road to rail in Britain and a much improved market share for British Rail (BR) with consequent reduction in lorry traffic on roads
4. The much-improved travel opportunities for European citizens with the possibility of frequent and fast journeys from many links in France, Germany, and Italy to important cities in Britain
5. The consequences for BR of improved freight and passenger revenue brought about by the exploitation of new business opportunities could mark a turn around in its fortunes
6. The Tunnel will almost certainly bring about a strengthening of European ties and European coherence, even if this is more symbolic than actual.

The Channel Tunnel idea

Proposals for a Channel Tunnel go back over 100 years and the project has come very close to fruition on more than one occasion; construction was actually begun in 1974 before the project was cancelled.

In 1985 the French and UK governments agreed to proceed with planning for a Tunnel and invited proposals for a fixed link catering for both road and rail traffic. Four schemes were submitted to the British government in October 1985 and subjected to an environmental appraisal (Land Use Consultant 1985). These four schemes are briefly described below, illustrating as they do the range of technical and economic possibilities under consideration at this time.

CHANNEL EXPRESS WAY SCHEME

This was originally a twin 11.3 m internal diameter bored tunnel arrangement connecting a UK terminal at Cheriton on the northern outskirts of Folkestone with a French terminal at Frethun, south-west of Calais (53 km). Each tunnel was to have been capable of accommodating both drive-through road traffic as well as trains but this was dropped and a new proposal made that both tunnels should be used exclusively for road traffic. A separate rail link would be constructed to run parallel with the road tunnels, connected by cross galleries.

CHANNEL TUNNEL SCHEME

This is the proposal eventually adopted for a rail-only link utilizing a 'shuttle' train service carrying road vehicles and foot passengers interspersed with railway freight and passenger services. The link would take the form of two 7.3 m internal diameter service tunnels, one each side of a 4.5 m–diameter service tunnel. The overall length of the tunnels would be 49 km between Castle Hill at the eastern end of the Cheriton terminal and Sangatte in France. The running tunnels are sized to accommodate specially constructed single- and double-decker 'shuttle' trains carrying lorries, coaches, and cars as well as normal high-speed trains.

EUROBRIDGE SCHEME

The Eurobridge comprises a four-level, twelve-lane road suspension bridge with spans of up to 5 km and a separate bored 6-m-diameter rail

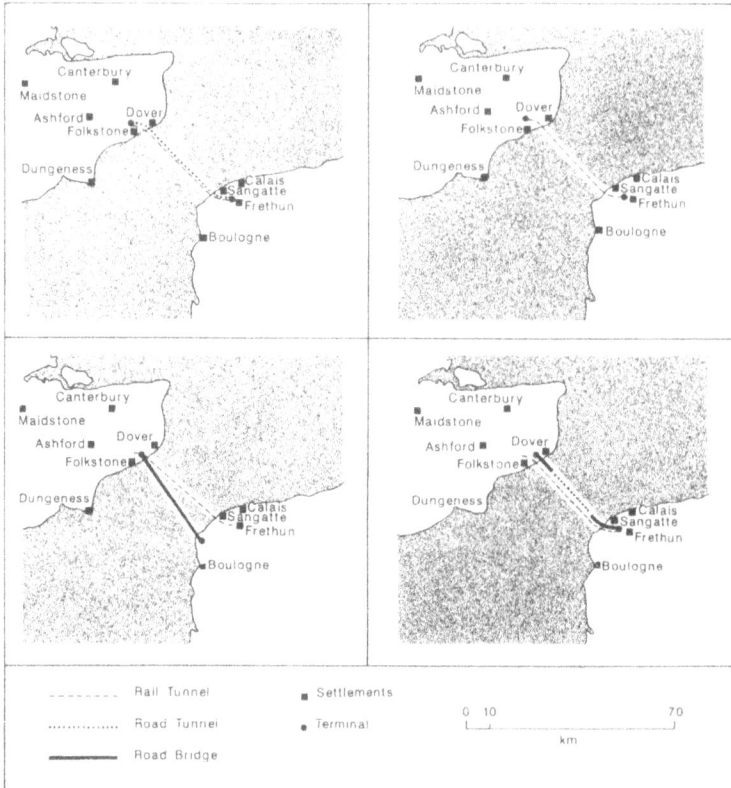

Figure 10.1 Routes to be taken by the four competing schemes for a Channel Tunnel

tunnel. The bridge decks would be enclosed and suspended 75 m above sea-level. The rail tunnel would be paralleled by a 4.5-m-diameter service tunnel connected to the tunnel by cross galleries every 500 m.

EUROROUTE SCHEME

The Euroroute scheme comprises a dual two-lane highway between a main UK terminal at Farthingloe, west of Dover, and a similar facility in France, south of Coquelles on the outskirts of Calais. The crossing to the Channel would be by cable-stayed bridges over inshore coastal waters and through a 21-km-long immersed tunnel under the main shipping lanes. The transition from bridge to tunnel would be made at artificial islands constructed clear of the main shipping lanes some 6.9 km from the shore. A rail link would be constructed at the same time by means of twin 6.2m–diameter bored tunnels. (The routes taken by the four schemes showing their landward terminals are illustrated in figure 10.1.)

DEVELOPMENT OF THE EUROTUNNEL SCHEME

On 12 February 1986 the British and French governments signed a treaty to agree to a Channel Tunnel scheme. A White Paper, *The Channel Fixed Link*, was presented to Parliament also in February 1986, advocating the Channel Tunnel Scheme, above, for an entirely rail-based solution. A

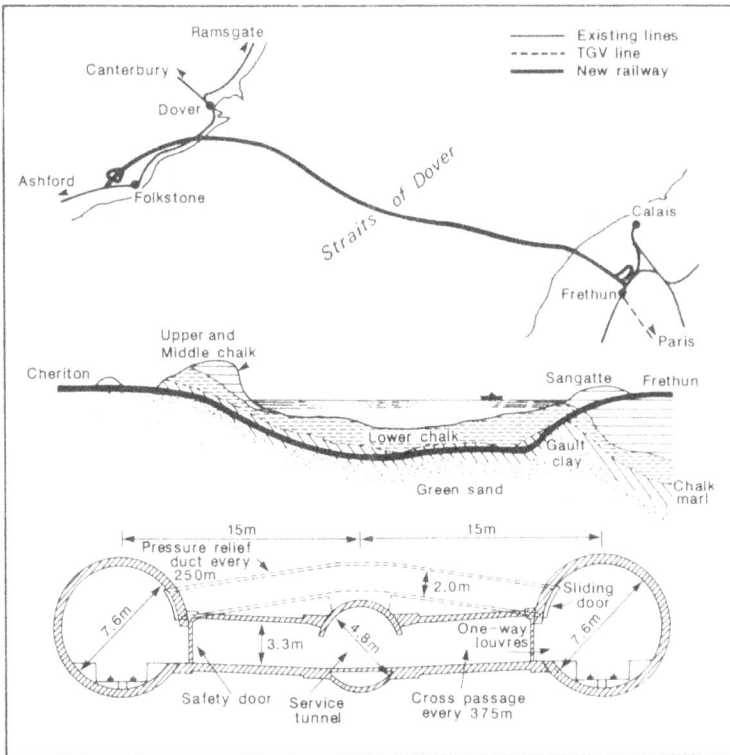

Figure 10.2 Details of the Channel Fixed Link

concession to finance, construct, and operate the Tunnel was signed in March 1986 with the proposer adopting the title of Eurotunnel. The main characteristics of this scheme are illustrated in figure 10.2.

The Channel Tunnel is to be financed entirely by the private sector with no government support. The capital cost of the project is £2,750 million, but with provision for overheads, inflation, and interest, this sum grows to £4,630 million. Eurotunnel propose to raise around £6,000 million to cover this amount plus contingencies. The European Investment Bank has agreed to a loan of £1 billion towards this total. The remainder will come from banks and other financial institutions.

In addition to this investment, there will be further substantial expenditure on the part of the railway companies. Expenditure by BR, SNCF and SNCB on track, railway stock, signalling, and other charges will amount to £2,087 million.

Rail developments associated with the Tunnel

It has become increasingly clear that the Tunnel cannot have the impact intended or generate rates of return which are acceptable to financial backers unless investment in the Tunnel itself is matched by landward-side investments. There are many different elements to be considered within an overall Tunnel investment strategy many of which are poorly developed at the present time. Table 10.1 lists some of the different elements of an overall investment strategy. The table does not consider rolling stock requirements.

At the time of writing, there are well-developed proposals for TGV running to the French terminal at Frethun but no proposals at all for a new line from the British portal to London. Similarly, there are no

Table 10.1 The Channel Tunnel: overall investment strategy

1. The Tunnel
2. Terminal facilities at both landfalls
3. Customs clearance facilities and freight handling facilities at both landfalls
4. High-quality rail links to London and Paris
5. Inland customs clearance facilities in Britain and France and further afield
6. International passenger journey facilities from areas in northern England through to France, Germany, and Italy, avoiding a transfer between stations in London; and equivalent possibilities in France
7. Closely related to (6), investment in cross-London facilities

proposals for inland customs clearance facilities in the UK which would go some way to relieving congestion in south-eastern England and transmitting the benefits of the Tunnel investment to northern England. The disparity between UK and mainland European preparation for exploiting Tunnel opportunities can be clearly seen in figure 10.3.

In the figure there are a large number of route possibilities utilizing a high-speed network and incorporating the Tunnel. Between the Tunnel and London, and around London, the picture is much bleaker with very

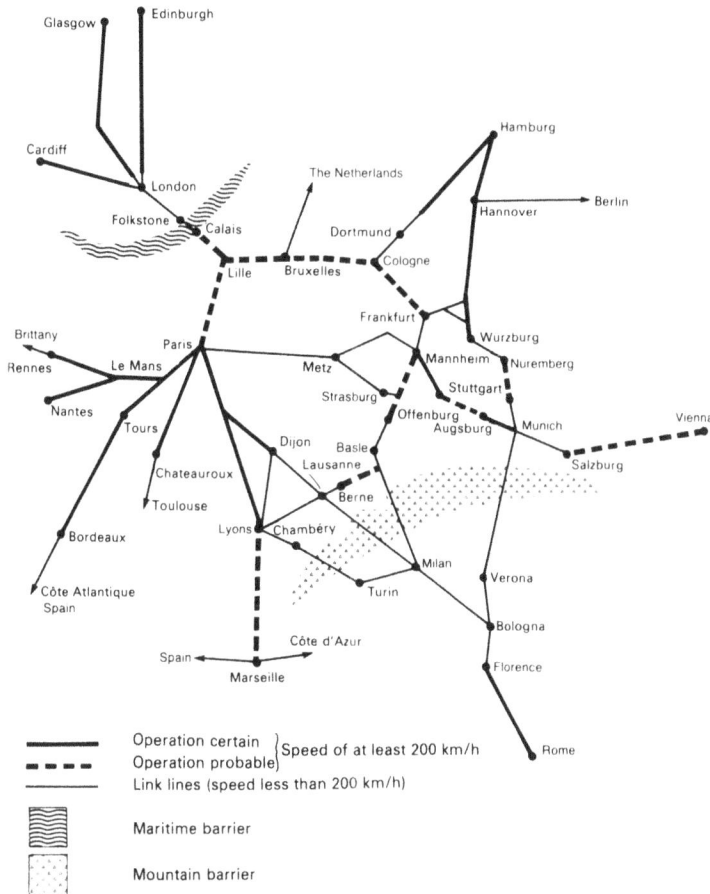

Figure 10.3 Map of rail networks and their links with the Tunnel

little improvement to existing track configurations and networks. The French are working on the high-speed line from Paris to Lille and the Tunnel portal as part of the Paris–Brussels–Cologne scheme.

On the UK side there will be no matching of the French investment in TGV-Nord. Upgrading the existing line is to be 'modest' to permit a maximum speed of 160 km/h compared with 145 km/h at present. BR's Tunnel-related investments amount to £450 million with large sums devoted to works at Waterloo Station (£45 million) and the North Pole Depot in north London (£42 million). A conference of local authorities in January 1987 expressed great dissatisfaction with this relatively poor level of investment and specified a series of measures which would spread the benefits of the Tunnel throughout the UK and improve the ability of the UK economy to capitalize on the investment. The recommendations of the conference included:

1. Upgrade and electrify to inter-city standards the link between Snow Hill, Farringdon, and King's Cross in London and its connections to the East Coast mainline at King's Cross and the Southern Region line at Blackfriars (without prejudice to existing and proposed local and regional services)
2. Upgrade and electrify to inter-city standards the West London line and its connections to the West Coast mainline at Willesden and the Channel Tunnel route at Clapham
3. Upgrade the Tonbridge–Redhill–Reading line, including a new link at Redhill to provide a direct through route
4. Bring forward proposals to electrify the Midland mainline between London and Leicester, Derby, Nottingham, and Sheffield; and the London to Bristol and South Wales mainline
5. Bring forward a scheme to operate frequent, direct passenger trains between the Continent and Coventry, Birmingham, Wolverhampton, Liverpool, Manchester, Glasgow, Peterborough, Doncaster, Leeds, Bradford, York, Teesside, Newcastle upon Tyne, and Edinburgh, and such other places as are considered appropriate
6. Provide and maintain suitable rail freight depots for customs clearance in South Wales, the West Midlands, the East Midlands, Merseyside, Greater Manchester, South Yorkshire, West Yorkshire, Teesside, Tyneside, and Scotland, and such other places as are considered appropriate.

Clearly a major problem with the Tunnel idea is a lack of investment away from the Tunnel portal and a marked lack of provision for maximizing freight and passenger services with origins in northern England and Scotland and destinations in mainland Europe.

Jonathan Roberts (1987) has discussed the pattern of services through the Tunnel. Waterloo Station will see up to four trains arriving and departing per hour in busy daytime periods, while up to twelve trains per day each way are proposed between the West Coast mainline and Paris/Brussels (six each way on the East Coast mainline). These services are subject to customs clearance on board, something which the British customs have been reticent to introduce. Under present plans passengers from South Wales, the West Country and the Midland mainline will have no through Channel Tunnel passenger service because there is no inter-city electrification planned for these routes.

Traffic forecasts

Eurotunnel have produced forecasts of cross channel traffic (figure 10.4). The forecasts show an immediate (1993) market share of 30 million passengers and 14 million tonnes of freight (8 million on trains and 6 million on lorries carried by shuttle). Roughly half of the 30 million passengers a year anticipated on opening will travel on the shuttle in cars and coaches and the other half in through trains. Estimates of how much freight and passenger traffic will be carried through the Tunnel should be treated with caution. The freight figures will be sensitive to the availability of services through the Tunnel which enable private sidings and inland customs clearance depots to be linked throughout Europe. With specific investment in a northern inland depot, as is suggested for Tinsley, near Sheffield, freight through the Tunnel could increase dramatically and continue to grow if several such depots serving major regions were to be established.

Steer, Davies, and Gleave (1987) identify five sources of uncertainty which have an influence on the market share of freight which the Tunnel might eventually achieve:

1. The pricing policy of the Tunnel operator, including that for the shuttle trains
2. The competitive reactions of the existing ferry operators
3. The pricing and institutional arrangements pursued by BR and the European rail administrations
4. The clearing arrangements agreed by customs authorities

5.   The actions of the respective railway undertakings in winning new business.

Steer, Davies, and Gleave adopt two main options in forecasting potential diversion to through rail freight.

Option 1 is based on conventional technology and an assessment of several different forecasts. It results in a forecast for 1993 for new rail freight of between 3.1 and 5.8 million tonnes. A growth rate of 3.5 per cent per annum is assumed beyond that point.

Option 2 incorporates the possibilities which arise from combined transport (see Chapter 9), which make for easier transfer between road and rail, but necessitate considerable amounts of new investment over and above what is contemplated in the UK. If combined transport could be exploited Steer, Davies, and Gleave estimate a potential transfer of 13–15 million tonnes of divertable traffic.

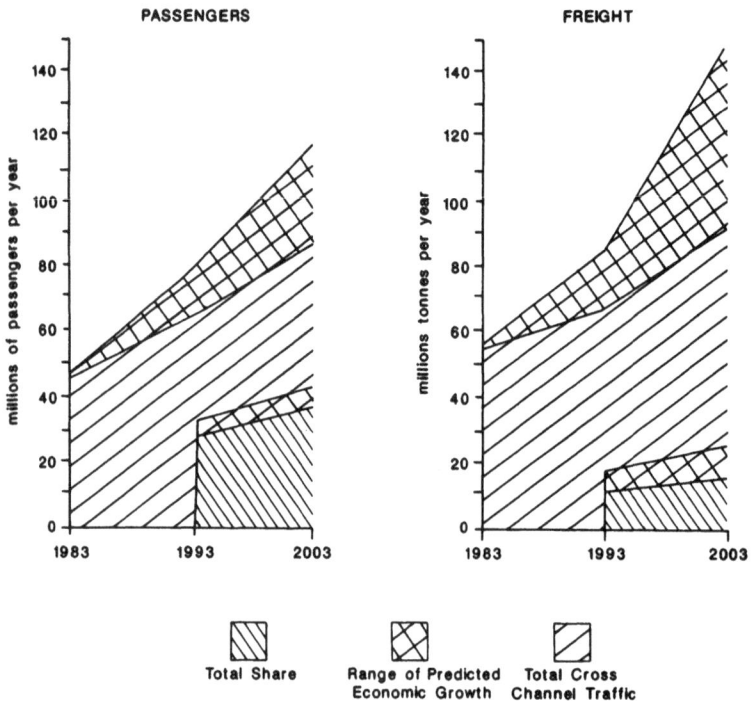

Figure 10.4 Forecasts of cross–channel traffic

Eurotunnel forecast that the Tunnel will capture only about 20 per cent of the total UK–Continent road freight (30 per cent of that on the short sea routes). Simmons (1987) makes the point that the projected growth in freight traffic overall could still leave the ports with more traffic than now even with the Tunnel in operation.

Channel Tunnel impacts

The Tunnel is a major transport investment which can be expected to have a large number of ramifying effects which range from regional development issues in the north and south of England through job loss in Kent, congestion in London and environmental improvements which may well result from a growth in rail freight and a transfer from road to rail.

There are problems with demonstrating the benefits of transport improvements in regional development terms. Whitelegg (1985) has reviewed some of the evidence in this subject, as has Simon (1987), and in both cases there is reason to doubt the job creation or economic revitaliz-ation effects of transport investments. Simon's conclusion is particularly relevant:

> notwithstanding the impact of recession, these results would seem tentatively to support other recent findings that regional development benefits from such projects are smaller than generally assumed ... county and regional employment rates are now worse, both in themselves and in relation to the national average, than at any time since at least 1971. (p. 34)

The Channel Tunnel and its impacts must, therefore, be treated with some caution. At the very least, there will be a range of positive and negative effects which must not be assumed to be self-cancelling. There will be environmental losses in Kent arising from road-based traffic heading for the Tunnel, environmental problems in London arising from the decision to concentrate so much traffic on Waterloo, and economic gains in places like Ashford and economic losses in places like Dover. In France it is likely that regional development effects will bring positive results to the region around the Tunnel entrance partly because of existing regional development incentives to a disadvantaged region, Nord–Pas de Calais, and partly because of greater levels of investment in related activities such as the TGV. It is quite possible that any gains made in this region of France may be at the expense of the relatively 'over-developed' south-east of England with its poorer level of rail infra-structure and exclusion from regional development incentives.

At a larger geographical scale, and larger time scale, we can expect the pattern of gains and losses on both sides of the Channel to become more complex. In particular, there will be waves of development which spread outwards away from the focus of the Tunnel portals. Aided by railway infrastructure investment and investment in combined transport, these benefits will rapidly spread from the French portal to Lille, Brussels, and Cologne, and along the route of TGV-Atlantique towards Nantes and Bordeaux and the TGV-Paris-Lyons. There are formidable forces working towards economic concentration on the European mainland and away from south-eastern England.

The final balance of economic advantage will depend upon the activities of at least three different sectors and the degree to which they can co-ordinate their activities.

Government who will be mainly responsible for regional policy incentives and the framework within which BR can raise cash to fund investment and private firms can move into rail freight through private siding operations. Government will also control important policies which affect arrangements for customs clearance and the possibilities of establishing inland customs clearance depots.

Railways who will need to make decisions on investment and make them quickly if opportunities are not to be lost. Rail links to London, electrifi-cation, and improvements to the UK loading gauge are urgent matters which need attention. Also important is the attitude of railways to the levels of freight throughput which are deemed acceptable for inland customs clearance operations.

Private firms will need to re-think their operations including shipping and freight movement to produce the best strategy for capitalizing on the Tunnel and its much reduced journey times for freight. Decisions on private sidings will be a high priority for large movers of freight.

All these opportunities and strategies are available to some extent to organizations in mainland Europe, many of whom are already well adapted to the market opportunities provided by rapid international transport. Further liberalization of international trade in Europe, regional aid to northern France, and well-funded combined transport operations in Germany and France could well lead to a situation where the Tunnel functions as an economic drain with the direction of flow being clearly outward. Vickerman (1987) poses this same dilemma in another way: 'There is no immediate advantage to be gained for any one UK region nor indeed for UK regions as a whole relative to other European regions ... any improvements in accessibility will have to be relative to that gained by other competitor regions to be significant' (Vickerman 1987:189).

If Vickerman is correct, turning accessibility benefits into regional economic benefits is going to be a highly competitive struggle with a large number of regions elsewhere in Europe. We will now examine three specific examples of Tunnel impact: environmental (including road to rail transfer); the impact on jobs; and the impact on London.

ENVIRONMENTAL IMPACT

Detailed environmental impact assessments of the original runners for the Channel link have been evaluated by Land Use Consultants (1985). These cover important matters such as job loss, dispersal of spoil, construction traffic, noise, air pollution, and so on. These are not the subject of this section, which focuses on more general issues than those relating specifically to Kent.

These broader issues have been succinctly demonstrated by Steer, Davies, and Gleave (1987), who claim a number of environmental plusses for the rail tunnel:

1. This scheme (unlike some others) will not damage the marine environment
2. Proximity to the M20 of the chosen site will avoid the need for major new road construction
3. A reduction of 888–1,044 heavy goods vehicles per day when the Tunnel opens equivalent to 6–11 per cent of the current number of heavy lorries on the M25 near the Dartford Tunnel (option 1 assumption in the Steer, Davies, and Gleave report)
4. A reduction of over 3,500 HGV movements per day, equivalent to 20–24 per cent of the M25 flow at Dartford if option 2 assumptions in the Steer, Davies, and Gleave forecast materialize.

They point out that the level of benefit gained if option 2 forecasts materialize is heavily dependent on improvements to the UK loading gauge which limit the size of wagons which can use the British Rail system. The continental gauge permits larger sizes to operate safely.

The report of Steer, Davies, and Gleave then translates the reduction of numbers of lorries into specific environmental improvements:

1. Reduction in noise levels 'on a typical stretch of road affected' on the M1 between junctions 9 and 10 ranged from 0.08 to 0.32 dB(A)
2. Reduction in air pollution again on the M1 between junctions 9 and 10. Carbon monoxide will fall from 9.6 p.p.m. to between 9.54 and 9.39 p.p.m. The levels of hydrocarbon will fall from 9.18 p.p.m. to between 9.15 and 9.07 p.p.m. The nitrogen oxides levels will fall from 1.54 p.p.m. to between 1.53 and 1.51 p.p.m.
3. Annual reduction in the number of fatalities will be in the range 2–6; a fall in serious injuries of something in the range 25–97, giving in total a monetary benefit of between £0.6 million and £1.88 million in 1993.

Eurotunnel are more optimistic in their glossy literature. They claim that the construction will create over 50,000 man-years of employment in both the UK and France. They also claim that £1.4 billion in orders will be placed for civil engineering and construction materials with further employment for Britain's principal industrial regions. Eurotunnel claim that 5,500 jobs will be created by the operational needs of the Tunnel once it is open, including 1,400 who will be needed to operate BR's new passenger and freight services between Britain and the Continent. None of these estimates attempt to tackle the question of jobs newly generated by the boost to competitiveness or economic efficiency provided by the Tunnel. Jobs in service industries in tourism, and in northern England, may well result from this boost but it would be a difficult task indeed to provide estimates of job creation and to set these against job losses in different sectors of the economy and different areas of the south-east.

One of the main concerns in the debate about the Tunnel has been the prospect of job loss in ports and on ro-ro ferries. Ferry-related employment in Kent in 1985 totalled 12,450, with 10,000 of these jobs at Dover (Channel Tunnel Joint Consultative Committee 1986). In the

Table 10.2 Ferry-related employment in Kent, 1985–2003

	1985	1993	2003
No Tunnel	12,450	13,200–13,800	14,000–15,400
With Tunnel	12,450	6,600– 9,500	7,400–11,300
Reduction before allowing for Tunnel employment		6,600– 4,300	6,000– 4,100

absence of the Tunnel, this total would be expected to increase to 13,200–13,800 (1993) and 14,000–15,400 (2003). On various assumptions of market share won by the Tunnel and trade lost by ferry operators, the conclusions reached by the Joint Consultative Committee are given in table 10.2.

Other factors considered include road damage and energy savings. The significance of Steer, Davies, and Gleave's (1987) results is not to be found in the quantification. The numerical results are not all that convincing, and are not well specified in terms of where the savings will be made and how they will vary over time. This is, of course, a difficult research problem and Steer, Davies, and Gleave are more than convincing on the principle that a major rail investment does bring with it a range of environmental improvements.

JOB CREATION AND JOB DESTRUCTION

During construction the Channel Tunnel is expected to require some 19,000 man-years of employment in Kent with a workforce peak of around 5,000 in 1990. Shortly after the opening of the Tunnel, it is anticipated that there would be 4,300–6,600 fewer ferry-related jobs in Kent, the largest reduction being at Dover (Channel Tunnel Joint Consultative Committee 1986). These reductions would be partly offset by 3,160 new jobs created in or associated with the Tunnel, mainly at Folkestone, Ashford, and Dover. The net effect would be a reduction of 1,400–4,000 jobs in eastern Kent around 1993.

A multiplier of 1.3 is assumed which produces an additional 30 jobs lost in the local economy for every 100 jobs lost on the ferries. To offset these job losses, at least in part, there will be jobs created in Tunnel operations and related activities. These are summarized in table 10.3. If job losses, job gains, and the multiplier effect are all put together, we see the picture presented in table 10.4. The final figure indicates a net reduction of 1,400

Table 10.3 Tunnel-related employment, including BR, 1993 and 2003

	1993	2003
Dover, Folkestone and Hythe	2,050	2,500
Ashford	1,110	1,210
Totals	3,160	3,710

Table 10.4 Estimates of port, ferry, and Tunnel employment in Kent, 1985-2003

	1985	1993	2003
No Tunnel	12,450	13,200-13,800	14,000-15,400
With Tunnel	12,450	9,800-12,700	11,100-15,000
Net change with Tunnel		−3,400 to −1,000	−2,900 to −400
Effect of employment multiplier		−1,000 to −300	−900 to −100
Overall change in employment with Tunnel		−4,400 to −1,400	−3,800 to −500

− 4,400 jobs brought about by the opening of the tunnel.

At present a consideration of a fully comprehensive job audit is out of the question. So much depends on forecasts of market share captured by the Tunnel, and of investment both near the Tunnel and up to 500 miles away, that it would be foolish to speculate. It may be difficult enough to consider employment spin-offs in Ashford or Calais, but it is more difficult to get anywhere near to orders of magnitude for the effects in Manchester, Sheffield, Lille, or Nantes.

LONDON

It is necessary to consider the effects on London because of the decision to concentrate most international passenger traffic on Waterloo Station and to make do with relatively poor-quality rail links in and around London which limits the scope for through travel from the north of England to continental destinations. This is not the case in France, with the TGV services virtually connecting with the terminal at Frethun.

The concentration of services at Waterloo is severely criticized by the London Strategic Policy Unit (LSPU) (1987) and has arisen because of customs requirements that there should be a traditional ferry port or airport customs arrangement rather than on-train examination, as is the norm on the Continent. On-train examination has subsequently been agreed but centralization at Waterloo with a customs hall is still the central feature of the London side of the operation. Undoubtedly, BR's operational requirements are very much simplified by one international terminal at Waterloo, supported by international facilities at Ashford.

1993   **2003**

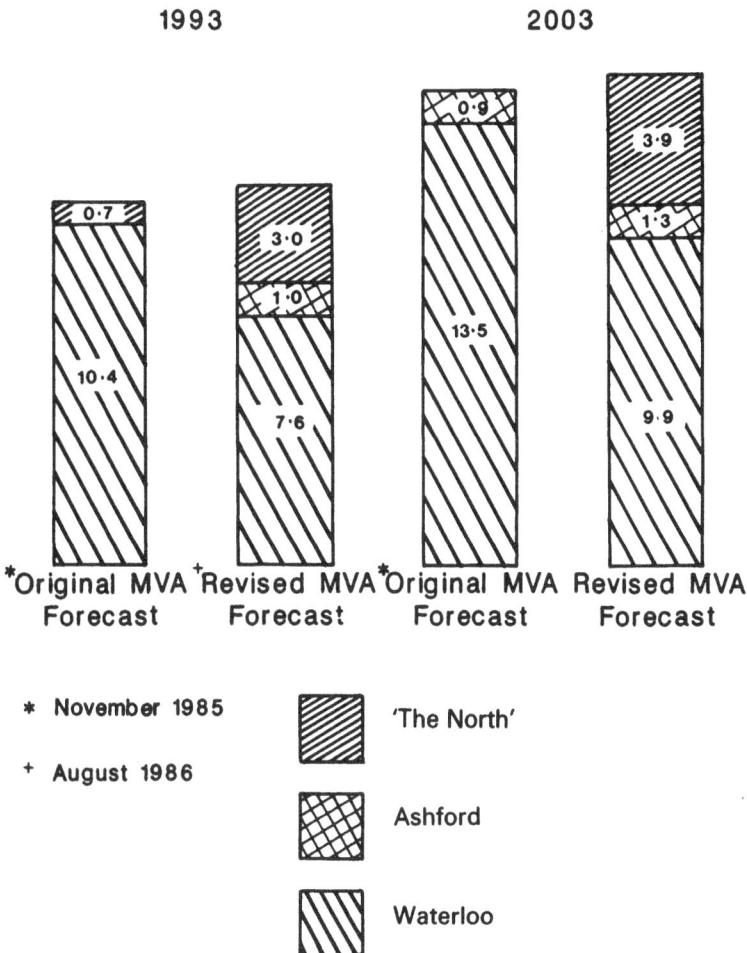

*Original MVA +Revised MVA *Original MVA Revised MVA
  Forecast   Forecast   Forecast   Forecast

* November 1985   'The North'

+ August 1986

                          Ashford

                          Waterloo

Figure 10.5 Forecasts of rail passenger journeys through the Tunnel with respect to UK origins of the traffic.

Forecasts of single journeys through Waterloo of approximately 10 million in 1993, and 13.5 million in 2003, have been made but these have also been criticized by the LSPU for their neglect of traffic generation effects. The consequences of this loading on road and public transport systems in central London are likely to be severe, while paradoxically they reflect a lack of provision of international facilities elsewhere in Britain. Figure 10.5 shows the forecasts which have been made of rail passenger journeys through the Tunnel and their geographic split in Britain.

The forecasts in figure 10.5 have not been justified by analysis of wider effects of such a burden in central London, nor of the possibilities for improving 'the North' figures if more than the twelve trains per day, each way, were introduced. The lack of a denser service pattern to the North and the complete absence of services to places like Sheffield, Nottingham, Leicester, Bradford, Hull, Teesside, Bristol, Cardiff, and Plymouth, all having populations over 250,000 will bring about inconvenient transfers between London termini and produce overloading on the underground system. The LSPU summarize the effect on public transport as follows:

1. Train capacity problems on underground services, accentuated by the dominantly north-bound flow of passengers alighting from Channel Tunnel trains and by the large number of passengers travelling with heavy luggage
2. Platform capacity problems at Waterloo underground station caused by heavy peak arrivals of passengers for departing Channel Tunnel trains
3. Terminal capacity problems and congestion on interchange routes, particularly escalators and tunnels from Channel Tunnel platforms to London underground services
4. Platform capacity problems at Waterloo Station caused by the conversion of platforms used for suburban services to Channel Tunnel train only platforms; this could mean the displacement of suburban services from Waterloo, rescheduling, or delays
5. Train capacity problems on the BR network in south London caused by the introduction of Channel Tunnel trains; this could result in the re-routing or even withdrawal of suburban services as well as the failure to implement useful new schemes
6. Overcrowding on bus services caused by peak arrivals of Channel Tunnel passengers, also poor reliability of buses caused by the increase in other road traffic; inefficient use of bus lanes because of illegal parking, and the large number of taxis, will also damage bus services around Waterloo, having spin-off effects further afield.

The LSPU describe in detail the likely outcome of heavy use of Waterloo Station; 40 to 50 per cent of Channel Tunnel passengers will transfer to the underground and, in turn, 40 per cent of these will patronize the Northern line (north-bound) and another 30 per cent the Bakerloo line (north-bound). Much of the traffic will be destined for mainline termini north of the Thames and will grossly overload the system. The LSPU describes this possibility as 'alarming even at today's demand figures'.

Conclusion

The Channel Tunnel is of fundamental importance to improved transport links in Europe, European integration, the removal of bottle-necks, and the success of the high-speed railway which is gathering momentum in France and Germany. With EEC proposals for eliminating border controls by 1992, and with the progressive liberalization of transport, the prospects for a brighter rail future in Europe have never been better. These prospects may well be matched by fundamental shifts in the structure of international trade between western and eastern Europe, Africa, and container traffic globally. Changes of such importance in the facilities for handling freight will bring about their own changes in the relative fortunes of the regions of the Community. These are too complex to be anticipated, but it is nevertheless clear that a new regional focus will emerge on the twin pivots of the Tunnel portals and a hinterland stretching inland from these points.

The UK landward side of the Tunnel planning has not come to terms with the possibilities and leaves large gaps in investment which will hamper for many years the exploitation of these possibilities. Similarly, northern England may well be bypassed by the same circumstances, depriving those areas of economic gain and contributing to the problems of congestion and environmental pressure in the south-east.

The Tunnel represents a major opportunity for the EEC to achieve its goals; for the UK to redress some of the economic decline; and for railways to recover a respectable market share of both freight and passenger business. At the time of writing (summer 1987), the balance is tipped against the possibility of the UK reaping these advantages to the full.

11 The environment

It is not possible to deal with environmental issues effectively as a self-contained, compartmentalized topic. At one level of interpretation, the whole of this book is about environmental issues. The effects of motorization on urban and rural communities, the effects on air and sound quality of heavy goods vehicles, and the enormous land-take consequences of motorway construction are significant environmental issues. We have already considered the pedestrian and the cyclist, and the importance of detailed urban design, which have been environmental in substance, and the significance of combined transport has as much to do with its environmental benefits as with reduced costs and better use of infrastructure.

There are wider issues still that are not considered in detail in this book. Motorized transport is heavily dependent on fossil fuels, particularly oil, whose exploitation poses environmental problems in addition to those caused by the partially burnt residues of hydrocarbon combustion. Dependence on higher and higher levels of motorization is inextricably linked with a tendency to dispersion of activities, their concentration in larger units with the 'inevitable' decline of public transport. These tendencies, taken together, seriously weaken the financial base of public transport (which is more energy efficient) and stimulate a considerable demand for new land uses, often on the edges of cities and based on car access. These interrelated land use and energy issues are discussed in Whitelegg (1979).

It is very difficult to draw a precise boundary around the environmental effects of traffic and transport systems. Air pollution figures significantly in the literature, as does noise and, to a lesser extent, visual pollution. These are also related to health effects. Rylander, Ahrlin, and Westerlund (1977) discuss the effects of noise on human health, and Blumer and Reich (1980) examine the effects of exhaust emissions on cancers. The study of human health problems and traffic pollution is not

well advanced and is complicated by stress-related effects in neighbourhoods which experience traffic.

There is a growing awareness that traffic and transport systems can
cause fear and anxiety, particularly in vulnerable groups such as women.
Women waiting at bus stops, travelling on the underground, or walking
along badly lit streets can be the object of sexual violence and attack. If
the chances of such assaults can be reduced by better design in the
transport environment, then there is a need to develop such approaches.
Road traffic accidents (RTAs) are another form of violence associated
with motorization. We have seen in chapter 6 how particular kinds of
environment are more conducive to RTAs than others. Fear and anxiety
can be induced by the prospect of crossing busy roads, competing with
inadequate timings on pedestrian crossings, and living with the heavy
lorry. New roads in the urban environment can slice through well-
defined communities and cause severe damage to walking journeys,
community cohesion, and the quality of life of residents. Community
severance on a large scale can still be found in British land use planning,
as in the Lancaster Local Plan, years after the damaging effects of such
severance have been identified (Lancaster City Council 1987). In this
chapter we will concentrate on the physical aspects of environmental
disturbance, particularly air pollution, the heavy goods vehicle, and road
construction.

Air pollution

The main pollutants associated with vehicles are carbon monoxide (CO),
nitrogen oxide (NO_x), photochemical oxidants, lead, hydrocarbons, and
other organic compounds such as aldehydes.

Carbon monoxide is a poison and decreases the oxygen-carrying capacity of
the blood. Carbon monoxide concentrations of 30 p.p.m. for 8–12 h have
been associated with impaired psychomotor performance and reduced
visual acuity, and with increased physiological stress to patients with
heart disease.

Hydrocarbons are a large group of organic pollutants, including poly-
nuclear aromatic hydrocarbons (PAHs), which are carcinogenic.

Nitrogen oxides react chemically with hydrocarbons to develop photochemical oxidants. Nitrogen dioxide is associated with increased
incidence of acute bronchitis in infants and young children, and acute
respiratory disease in the entire family group. Increases in the incidence

Table 11.1 Estimated pollutant emissions following combustion of fossil fuels in London, 1978 (thousand tonnes per annum)

	Carbon monoxide	Sulphur dioxide	Nitrogen oxides	Smoke	Hydro-carbons	Lead
Vehicles						
Petrol	900	2.2	37	3.4	44	0.8
Diesel	20	3.1	16	3.1	3.5	–
Sub-total	920	5.3	53	6.5	47	0.8
Domestic						
Solid fuel	24	8.6	0.8	6.0	1.6	–
Oil	0.2	1.0	0.7	0.1	–	–
Gas	0.9	–	3.6	–	0.4	–
Sub-total	25.1	9.6	5.1	6.1	2.0	–
Commercial and Industrial						
Solid fuel	0.4	10.8	3.0	0.4	0.2	–
Oil	1.6	80	13.9	1.4	0.3	–
Gas	0.6	–	5.7	–	0.2	–
Sub-total	2.6	90.8	22.6	1.8	0.7	–
Total	950	105	81	14	50	0.8

Note: Power station emissions are not included in this table. Rounding errors may be present.
Source: GLC (1983).

of respiratory disease have been associated with nitrogen dioxide levels ranging from 0.06 p.p.m. to 0.08 p.p.m. over a six-month period. In the UK motor vehicles contribute about 30 per cent to total NO_x levels.

Photochemical oxidants result from a complex series of atmospheric reactions initiated by sunlight, e.g. ozone and peroxyacl nitrate. These are injurious to human health.

Lead is a significant neurotoxin with particularly damaging effects on children and on the intelligence of children. It accumulates in persons who are exposed and the main source of lead is the petrol-engined motor vehicle.

Table 11.1 shows estimated pollutant emissions in London and presents the proportion which is the responsibility of vehicles. While the data in

Table 11.2 Pollution emissions, by source, 1984 (thousand tonnes)

	NO_x	CO	HC	SO_2
Domestic	48	370	56	160
Commercial/public service	44	11	1	140
Power stations	621	52	10	2,500
Refineries	37	4	1	150
Agriculture (fuel use)	3	–	–	10
Other industry (fuel use)	173	65	3	520
Rail transport	35	13	9	10
Road transport	716	4,456	538	40
Incineration/agricultural burning	12	220	38	–
Gas leakages	–	–	384	–
Industrial processes and solvent evaporation	–	–	600	–
Road transport (%)	42	86	33	1

Source: Holman (1987).

table 11.1 refers to one large metropolitan area, it is reliable as an indicator of the scale of pollution in many large cities across Europe, where people and cars are concentrated in relatively small geographical areas; it is compared with table 11.2, covering the UK as a whole, in 1984.

The situation in London is well documented (GLC 1983). Petrol-engined motor vehicles account for the vast majority of carbon monoxide emissions into London's atmosphere, amounting to 900,000 tonnes annually. Measurements of air quality, close to heavily trafficked roads, show maximum 1 h means averaged over three months of between 18 and 24 mg/m³. Values as high as 60 mg/m³ have been recorded, even at sites remote from roads during high-pollution episodes. The GLC's standard for carbon monoxide is 10 mg/m³ (18 h mean).

The main contributor to airborne lead concentration in London is the petrol-engined motor vehicle. Table 11.3 shows some measures of lead content in street dust where the guideline standard is 500 µg/g lead in dust.

Airborne lead in London frequently exceeds the GLC's guideline of 1 mg/m³ (which compares with the EEC limit value of 2 mg/m³). The reduction of lead added to petrol in Britain (to 0.15 g/l) has contributed to a reduction in the severity of this problem, but not its elimination.

Table 11.3 Lead content of street dusts (in µg/g)

Area	Number of samples	Mean concentration of lead and standard deviation
Lambeth (central)	15	1,840 ± 510
Islington (central)	15	1,330 ± 660
Wimbledon (mid-urban)	16	1,040 ± 300
Edgware (mid-urban)	15	1,040 ± 450
Harold Wood (outer urban)	18	920 ± 500
Walton-on-the-Hill*	7	35 ± 15

*Footpath in a rural area.
Source: GLC (1983).

Table 11.4 Nitrogen dioxide concentrations measured at County Hall, London (in µg/m³)

	1978-9	1979-80	1980-1
Rooftop site 30 m high:			
Hourly mean			
Summer	48	39	45
Winter	46	49	50
Annual	47	44	47
99 per cent of hourly values			
Summer	143	91	117
Winter	133	109	145
Annual	139	103	135
Roadside site 7 m high:			
Hourly mean			
Summer	73	–	57
Winter	–	67	81
Annual	–	–	69
99 per cent of hourly values			
Summer	240	–	143
Winter	–	172	226
Annual	–	–	194

Source: GLC (1983).

Traffic is a major source of nitrogen oxides. In London a smaller amount comes from stationary emitters than from traffic. Nitrogen dioxide levels are recorded in table 11.4.

Air pollution from vehicles is related to the size of the vehicle population, its driving characteristics (e.g. idling, cruising, accelerating, decelerating) and hence the amount of congestion. Measures to eliminate pollution will be measures which can reduce the size of the vehicle population and its driving characteristics as well as those technical measures which might be applied to vehicles themselves to minimize pollution from exhaust and engine systems. Policies which encourage walking, cycling, public transport use, and diversion of freight to less polluting modes, such as railways or inland waterways, will all contribute to improved air quality in cities.

Noise pollution

Noise is more difficult to deal with as a pollutant than the chemical pollutants discussed above, but it is no less a serious problem. Motor vehicles are noisy – and some are more noisy than others. Motorways and lorries cause particularly serious disturbance. Physical measurements confirm that the intensity of traffic noise exceeds that of any other source over the greater part of urban areas (Gakenheimer 1978). A survey carried out in London, in 1961, showed that noise from motor vehicles predominated at 84 per cent of locations, a finding which has been reinforced by subsequent studies of noise nuisance (Yeowart, Wilcox, and Rossall 1977).

Noise affects people at several different levels. There is a physiological response which induces stress and may be related to loss of sleep or act independently as a general irritant. Gakenheimer (1978) refers to deafness among elderly people as something which may be accelerated by traffic noise, but for the largest group of sufferers, noise is simply an unwelcome and unpleasant distraction from normal everyday (and everynight) activities which seriously reduces quality of life on busy roads and in metropolitan areas generally.

The wider area effects are not well documented because of practical assessment difficulties, and there is very little documentation about the cumulative effect on urban neighbourhoods of noise, air pollution, road traffic accidents, fear, insecurity, and community severance. Taken together, and evaluated at neighbourhood level on an area-wide basis, the result would graphically display the enormity of the negative impact of motorization and the degree to which this effect goes unrecorded in any systematic fashion. One study of the London borough of Kensington and Chelsea goes some way to meeting this objective, but still concentrates on the 'hard' measurable parameters (Joint Unit for Research in the Urban Environment 1976).

Sound is measured on a logarithmic scale in decibels (dB) to provide a scale which roughly corresponds with the range detectable by the human ear. Sound-level measuring instruments usually apply the so-called 'A' frequency weightings and measurements are expressed in units dB(A). An increase or decrease of 10 dB(A) represents a doubling or halving of loudness. Thus a noise measured as 80 dB(A) will sound twice as loud as one which registers 70 dB(A). As this is all somewhat obscure, some examples of typical sound levels in dB(A) are given below:

120 discothèque - 1 m from front of loudspeaker
100 pneumatic drill at 5 m
 90 heavy goods vehicle from pavement
 80 powered lawnmower at operator's ear
 70 vacuum cleaner at 3 m
 50 boiling electric kettle
 40 refrigerator humming at 2 m
  0 total silence.

It should be appreciated that sound levels will vary enormously in an urban area with the three-dimensional geometry of the urban form as well as the number and type of the vehicle population. Under these circumstances motor vehicles and motor-cycles, which are manufactured to particular noise limits, may well continue to cause disturbance and annoyance. In France a mean sound level of 60 dB(A) in front of residential property has been recommended as a limit not to be exceeded; US studies have shown that when daytime noise levels are greater than 66 dB(A), relaxed conversation is not possible.

Noise levels seem to be on the increase, though detailed data required to verify this view are difficult to collate. Gakenheimer (1978) refers to an increase in traffic noise levels of 2-3 dB(A) in France between 1970 and 1985. Spatial variations are likely to be most marked, and the impact of new roads, especially in urban areas, is a significant factor in determining the size of the population exposed to excessive noise.

Heavy goods vehicles

The heavy goods vehicle (HGV) is often seen as the major culprit in discussions of environmental problems arising from traffic (Baughan, Hedges, and Field 1983). The HGV is a conspicuous user of road space and is likely to be associated with more noise, danger, and pollution than the car. In the UK the issue has been the subject of a government enquiry (Armitage 1981), and an even more searching enquiry at the level of one large city - London (Wood 1983). Both enquiries have produced a mass of data on the impact of the HGV but have failed to resolve the argument, which basically revolves around the costs and benefits of lorry-borne freight, the issue of taxation and track costs, and the potential for switching freight from road to rail and from road on to a combined transport system.

Noise is one of the main problems experienced as a result of lorry operations. HGVs make between 88-92 dB(A) of noise, and light goods vehicles 79-81 dB(A). Lorries are noiser (by 9 dB) than cars in free-flow situations, but noisier again (by 17 dB) when the traffic is congested. Gradients and frequent stops and starts, as on a signalized street in an urban area, will make the lorry nuisance worse. The results of several surveys are discussed in Haigh and Hand (1983) which indicate the degree of annoyance felt by residents in different towns in response to the incursions of the lorry.

Research in Japan and the UK has suggested that the noise power emitted from vehicles exceeding 1.5 tonnes gross vehicle weight (GVW), travelling freely at a given speed in the range 40-100 km/h, rises by an average of about 0.1-0.2 dB(A) for each tonne increase in GVW (OECD 1983b). Articulated vehicles (tractor and semi-trailer with two axles) produce noise levels that average 2-3 dB(A) greater than a rigid truck.

Research has been conducted in Europe with the 'quiet heavy vehicle' (OECD 1983b), and the GLC before its abolition was encouraging the fitting on to HGVs of 'hush-kits'. It could be argued that these techno-logical approaches to a serious problem, while effective within limits, are no substitute for general policies on the movement of freight by whichever mode is environmentally more sensitive, and the careful planning of lorry routes and lorry bans as in the GLC lorry ban which followed the Wood enquiry.

Generally poor land use planning and badly located industrial estates can be the basic cause of lorry nuisance in urban areas, as in the case of the Lune Industrial Estate in Lancaster which necessitates hundreds of lorry movements per week around unsuitable residential roads. The structure of land use and transport planning in the UK is such that problems like this can rarely be solved by re-location which is the only remedy which would produce the desired outcome.

Noise is very closely allied with ground-borne vibration and the damaging effects of HGVs on buildings, bridges, and underground services such as gas mains and sewers. The UK Department of Transport estimates that 11,260 bridges need bringing up to strength, at a cost of £830 million, to cope with the additional demands of heavy traffic and 38-tonne HGVs (RoSPA 1987). Research by the US Federal Highway Administration (OECD 1983b) shows that ground vibration effects are mainly dependent on GVW and their speed. Ground vibration levels are roughly doubled for each doubling of GVW, but if the same gross weight is distributed over additional axles, less vibration is experienced.

Haigh and Hand (1983) quote several surveys of the degree of annoyance caused by vibration and found high levels of annoyance, both personally and through the effects on buildings, particularly homes. There is still considerable argument about the physical effects of vibration damage to buildings, less so with the damage caused to roads and bridges.

Lorries are implicated in air pollution, particularly in urban areas where driving conditions may lead to excessive smoke emissions. Diesel engines go some way to 'softening' the impact of the HGVs' air pollution, but lorries are still significant contributors to air pollution (see later in this chapter). In the case of smoke, lorries emit four times the quantity of particulates than do cars, and in a London survey of the 34 per cent of smoke attributable to motor vehicles, 80 per cent came from lorries (Armitage 1981).

Roads and the impact on the environment

Roads are big business. They are significant sources of contracts for private sector construction firms and they are major items of public expenditure. Table 11.5 shows public expenditure on highways between 1960 and 1980 in Germany. In 1984–5 expenditure in the UK was £2,300 million, comprising expenditure on motorways and trunk roads of £800 million and local authority expenditure on all other roads of £1,500 million.

In the UK roads are built to satisfy three main objectives:

1. to assist economic recovery by reducing transport costs
2. to improve the environment by removing through traffic (especially lorries) from unsuitable roads in towns and villages
3. to enhance road safety.

Serious doubts have been expressed about the first of these objectives (Whitelegg 1985), and the third does not hold up to close examination when increased traffic flow and area-wide (as opposed to individual stretches of road) accident statistics are taken into account.

The second objective is more difficult to assess. In many situations a bypass around a congested village or town will bring speedy relief from heavy traffic and lead to dramatic improvements in the centre of the town. In northern England, Appleby-in-Westmorland has clearly benefited from its bypass, while Settle still suffers from excessive traffic volumes (though a bypass is currently under construction). New roads, however, may be a mixed blessing. Not all traffic will be diverted and the proportion of through traffic may have been overestimated. If more is diverted, this may have a detrimental effect on the local economy. More-over, new roads will lead to more traffic being generated and eventually finding its way into the built-up area; and additionally, new roads do take up valuable land and detract from the visual quality of the landscape. These are difficult effects to quantify, and research into the effects of new road construction and additions to a rural or urban environment is not well advanced.

Methods of evaluating costs and benefits of new roads in the UK (mainly COBA) are weak on environmental considerations and place too much emphasis on time saving for motorists. In spite of this built-in bias towards new roads, COBA frequently returns a negative or low economic benefit and roads are still built. This situation does not apply in

Table 11.5 Public expenditure on highways (in million Deutschmarks)

	1960	1965	1970	1975	1980
Federal state	1,700	3,222	5,528	6,996	7,853
States	1,280	2,665	4,233	5,528	6,984
Local	2,745	3,820	5,132	5,037	8,232
Other	150	256	–	–	–
Total	5,875	9,963	14,893	17,561	23,069

the case of public transport, especially rail investment. A National Audit Office (NAO) investigation of economic assessment was critical of this approach to road construction. In 1980, 109 road schemes were dropped because of financial considerations:

> NAO analysis of the suspended schemes disclosed that in 56 cases no economic evaluation had been carried out at that time and there was no other quantified indication of the relative importance of these 56 suspended schemes. 19 were subsequently reintroduced into the programme published in the 1983 White Paper without any detailed economic evaluation having been made in the intervening 2 years.
> (NAO 1985: 8, para. 2.5)

The same NAO investigation showed schemes going ahead with negative returns (based, it was argued, on environmental grounds) and revealed that estimates of traffic which were forecast at the design stage were not being compared with actual flows, one year after construction, as they should have been. One-third of these schemes showed variations in excess of ± 20 per cent. The NAO investigation shows that, in road matters, the relevant government department has little idea of the environmental impact of new roads and continues to build roads in the absence of this information – and in the light of negative economic returns and inaccurate traffic forecasts.

With such a commitment to road construction, it is not surprising that there are so few studies of the impact of new road construction (and no studies at all of how far the government's objectives for road construction have actually been advanced by the road programme. One example of such a study is that by Secrett and Hodges (1986) on roads and their impact on the natural environment. They review the damage which is caused by road construction to the environment and found that 24 sites of special scientific interest (SSSIs) have been damaged and 38 SSSIs are threatened with serious damage, while 45 additional SSSIs will suffer some level of damage. They point out that SSSIs are the minimum amount of environmental protection that should be sustained and that this minimum threshold is regularly breached. The loss of habitat and species beneath roads and their associated works is particularly serious in the case of motorways, which consume approximately 26 acres per mile. New roads create a variety of environmental problems. If the water-table is covered by engineering works, then wet-land ecosystems may be destroyed. If roads split up woodland and farmland, the integrity of a particular habitat for plants and animals may be destroyed; this can lead to rearrangement of fields and loss of hedgerows. Roads as barriers lead to casualties to owls, butterflies, and toads, among a number of other vulnerable species. As partial compensation, roads can create new habitats, including the 18,000 acres verging Britain's 2,000 miles of motorway. Such new areas, while valuable in themselves, cannot be regarded as an acceptable trade-off, though they should figure in an environmental impact assessment if a formal system along US lines were introduced.

One of the most serious examples of road-based environmental destruction is the East London River Crossing proposal which would cut a swathe through the Oxleas Wood SSSI. Oxleas Wood is unique in London and has evolved since the end of the Ice Age to produce a complex and irreplaceable flora and fauna that would be valuable in any area, let alone in a huge city. Its incorporation into a major road scheme shows the very low implicit valuation which is placed on environmental goals in road construction schemes.

Salt applications on roads present particular problems for plant and animal species. In the severe winter of 1978-9, 12 tonnes per lane per kilometre of salt were applied to UK roads. Saline drainage affects the viability of freshwater flora and fauna, and salt spray can be taken up directly by plants up to 9 m from the carriageway.

The thread running through many of the environmental issues which are raised by road construction programmes is the relatively low priority accorded to environmental impact and the difficulty surrounding the mechanism of evaluation. Transport 2000 has reviewed the question of environmental evaluation as part of its evidence to the (UK) Standing Advisory Committee on Trunk Road Assessment (Transport 2000, 1985). It concluded that the Department of Transport's *Manual of Environmental Appraisal* needs substantial revision. Transport 2000 wants to see the opportunity cost of new roads incorporated into any assessment, so that the return on the road can be compared with returns from other land uses. Transport 2000 has specific criticisms of other aspects of the environmental assessment and find serious flaws in the methods and emphases attached to noise, air pollution, community severance, visual impact, heritage and conservation areas, ecological factors, and the implications for public transport users, residents, and cyclists.

EEC activity in transport and environmental matters

The European Community's environmental policy was founded in 1972. A large body of legislation has been developed on the basis of article 2 and more important, articles 100 and 235 of the Treaty of Rome, though this was not based on a formal legal competence in this area. This situation changed in 1985 when the Council of Ministers adopted the draft Single European Act, which amends the European Coal and Steel Community (ECSC), EEC, and the European Atomic Energy Community (EURATOM) treaties and makes significant changes both to the institutional structure of the EEC and to further policies, particularly environmental ones.

The First Environmental Action Programme of the EEC (1973–7) spelled out the objectives and principles of community environmental policy; the Second Environmental Action Programme (1977–81) updated and extended the first but the nature of the Third Programme was very different. The Third Programme (1982–6) declared that environmental policy is a structural policy that must be pursued steadily, despite economic fluctuations, and must increasingly adopt a preventative character, so that natural resources will not be despoiled. The Third Programme states that the concern for the environment must increasingly be integrated with the planning of all activities in agriculture, industry, tourism, and transport.

The Council has adopted about 100 European legal texts in the field of the environment but this activity has not been matched by obvious improvements to the environment (Klatte 1986). Environmental directives are frequently not implemented, or are implemented late. Klatte maintains that hardly anything has been accomplished in the field of integration of environmental considerations into other community policies.

The Commission's Fourth Environmental Action Programme runs from 1987 to 1992 and constitutes a definite move away from reacting to environmental problems after they have arisen towards a general preventative approach. The Fourth Programme was adopted by the EEC in March 1987 and specifies eight priority areas:

1. Further action to reduce air pollution
2. Intensified efforts to deal with water pollution
3. Strict controls on chemicals
4. New measures to govern biotechnology

5. Safe use of nuclear power
6. A new emphasis on conservation of natural resources
7. Protection of the soil
8. Disposal and recycling of waste.

Associated with the Fourth Programme is the European Year of the Environment (EYE) which commenced on 21 March 1987. The main aim of EYE is to raise awareness of the importance of environmental protection, so that progress can be made in conserving and improving the environment. It lists transport and the environment among its areas of interest, but in conception it sounds much like the European Road Safety Year, which provided little of lasting benefit to the improvement of road safety and was characterized by a series of 'prestigious' national and international conferences.

Both the Fourth Programme and EYE have failed to grasp the significance of motorization, roads, pollution, and the architectural and cultural heritage of urban areas as an environmental issue of great importance. No concrete proposals have emerged for getting to grips with substantial issues like conservation in historical towns or public transport support, or walking and cycling, all of which have made larger contributions to environmental improvement to the largest concentration of people in Europe than the listed items of the Fourth Programme. The absence of traffic (and its pervasive and cumulative effect on cities) from the list is a major flaw.

There is also a technical bias in the environmental statements which ignores the real social context within which traffic causes damage. Blight, dereliction, community severance, and noise adversely affect communities and threaten the bondings which actually hold European civilization together. If cities cannot be held together as pleasant places in which to live and work, then more than a thousand years of European history will be brought to an end as society evolves towards a technocratic, polluted, and fragmented end-point, sustained by electronic gadgetry and hyper-mobility but lacking substance as a community and as a civilization.

The Community's environmental programme is rooted in different concerns which amplify basic community imperatives:

1. To improve the living and working conditions of European citizens
2. To facilitate the smooth functioning of the Common Market. Different environmental standards and technologies would hinder the flow of consumer goods, particularly cars, and establish different

financial burdens on producers. The Community strives to avoid these market 'imperfections'

3. To tackle the international/trans-frontier diversion of pollution, particularly air pollution. The acid rain controversy is a good example of the need for international agreement and collaboration which the EEC could foster.

In the next section we will examine some particular facets of EEC environmental policies as they affect transport.

Lead in petrol and vehicle emissions

The EEC has been very active in the area of lead in petrol and vehicle emissions, and this activity has come under the scrutiny of the House of Lords Select Committee on the European Community (House of Lords 1985). Harmonized legislation for the control of vehicle exhaust emissions throughout Europe was first introduced following the adoption, in 1970, of the United Nations Economic Commission for Europe (UNECE) Regulation 15. This Regulation initially controlled only carbon monoxide (CO) and hydrocarbons (HC); this has been amended four times in subsequent years. EEC activity through its Directives is summarized in table 11.6.

In 1987 the Commission moved a step further towards the elimination of lead in petrol and reduction in emission levels. By 1 October 1989 the twelve member states must ensure the availability and balanced distribution of unleaded petrol with a motor octane number (MON) of 85.0 and a minimum research octane number (RON) of 95.0 at the pump ('premium').

Table 11.6 EEC environmental Directives on lead and vehicle exhaust emissions

70/220	sets maximum amounts of gaseous emissions of CO, HC, and NO_x
72/306	sets acceptable smoke levels from heavy duty vehicles
78/611	lays down lead content in petrol which must not be greater than 0.4 g/l and not less than 0.15 g/l
78/665	reduced acceptable limits for gaseous emissions
80/779	sets air-quality limit values for sulphur dioxide and suspended particulates
83/351	reduces the acceptable levels of gaseous emissions
85/203	sets air-quality limit values for nitrogen dioxide

Table 11.7 Summary of vehicle emission limits

Directive	Date in force	Emissions (grams per test)			
		CO	$NO_x + HC$	HC	NO_x
78/665	October 1981	96		15.6	11.2
83/351	October 1986	67	20.5	–	–
Stage 1	October 1991	45	15	–	6
Stage 2	October 1995	10–35	2.6–8.2	–	1.1–4

The distribution of unleaded pumps in October 1986 was very uneven: Germany had 10,000, the Netherlands 8,000, Denmark 900, the UK 146, Italy 90, and France 89. The 1987 Meeting of Environment Ministers noted that little progress had been made with its 1985 proposals for vehicle emission (table 11.7). At the 1987 meeting there was agreement to help Greece over pollution in Athens, but Denmark considered the step to be too weak and refused to co-operate.

Diesel vehicles

It is often assumed that diesel engines are environmentally more accept-able than petrol-driven engines. Diesel fuel is lead-free, but in the case of SO_2 emissions, diesel engines may produce four times as much per litre of fuel than petrol engines (Holman 1987). A pollution emission inven-tory in London suggests that 8 per cent of the emissions in Greater London are from traffic, of which 80 per cent is from diesel vehicles; diesel vehicles produce more NO_x than petrol vehicles. Using British emission factors, it has been estimated that diesel vehicles produce approximately 190,000 tonnes of NO_x per annum – about 25 per cent of the UK total from traffic (Holman 1987). Diesel engines emit little CO – 5 per cent of the UK total, compared with 81 per cent from petrol vehicles – and considerably less gaseous hydrocarbon than petrol engines. Of the 538,000 tonnes emitted by traffic each year, diesel vehicles account for 8 per cent, compared with 92 per cent from petrol vehicles. Diesel vehicles give a higher polyaromatic hydrocarbon (PAH) emission than petrol vehicles, and diesel vehicles have become a major source of particulate emission as black smoke. A diesel-fuelled vehicle emits ten times more particulates under urban driving conditions than a petrol vehicle. Table 11.8 summarizes emission factors for diesel and petrol engines.

Table 11.8 Emission factors (in grams per litre of fuel consumed)

	Diesel < 3.5 ton*		Diesel > 3.5 ton†		Petrol	
	urban	non-urban	urban	non-urban	urban	non-urban
CO	10	10	25	12	200	125
Total HC	10	3.5	11	5.5	35	20
SO$_2$	4.2	4.2	4.2	4.2	0.3	0.3
NO$_x$	10	12	45	55	17	40
Particulates	7	4	14	7	0.65	0.65
Aldehydes	0.8	0.25	0.8	0.4	0.7	0.4
PAH ($\times$ 1,000)	8	5	16	8	0.7	1

*Based on an indirect inject engine.
†Based on a direct inject engine.
Source: Holman (1987).

Table 11.9 Proposed EEC emission standards for heavy vehicles

Carbon monoxide	11.2 g/kWh	9.0 g/Hp.h
Hydrocarbons	2.4 g/kWh	1.7 g/Hp.h
Nitrogen oxides	14.4 g/kWh	10.6 g/Hp.h

The EEC has proposed a Directive for controlling particulate emissions from diesel cars which was submitted to the EEC's Environmental Council on 12 July 1986. The proposed limit for particulates is 1.3 g per test. Smoke from heavy-duty vehicles is controlled by Directive 72/306/ EEC, and while these standards are used by manufacturers, they are difficult to enforce on the road. Also in 1986 the Commission submitted to the Environmental Council proposals for setting limits on the emission of gaseous pollution from commercial diesel vehicles (table 11.9).

Motor-cycle noise

Noise is a serious environmental problem disturbing large numbers of people and is particularly associated with traffic – lorries, motor-cycles, aircraft, and to a lesser extent rail vehicles. Noise produced by motor-cycles is controlled by Directive 78/1015 for both environmental and economic reasons. It is important for the community manufacturing interest that some standards should apply in this area and that they should not inhibit Community-wide trade in motor-cycles.

Under Directive 78/1015, member states may not set noise limits more stringent than 86.0 dB(A) for motor-cycles of over 500 cc, with lower levels for smaller machines. This Directive provided for further reduction in noise by 1985 and the EEC now proposes new limits:

size (cc)	maximum permissible sound level dB(A)	
	(from 1.10.87)	(from 1.10.95)
80	77	75
80–175	80	78
175	82	80

If anything these reductions in noise represent little improvement and do not reduce noise levels as much as does Directive 84/424 for cars and lorries.

In a summary of EEC environmental policy (EEC 1987) it is claimed that community environmental policy is concentrated around two principal themes: (a) the fight against pollution and nuisances; and (b) improved management of land, of the environment, and of natural resources. The second category makes no mention of roads and their associated land-take as an environmental concern, or of roads in the discussion of flora and fauna. It is clear that EEC environmental policies are skewed towards the technological end of the spectrum, where the additional concern for unified markets and free movement of goods gives a degree of respectability to environmental matters. Where such market concerns have no immediate relevance, there are glaring omissions which produce an inadequate response to the environmental threats posed by higher levels of motorization and greater degrees of societal dispersion.

The EEC has been active in an area of environmental legislation which could conceivably remedy these defects. The EEC Directive 85/337 introduces the idea of environmental impact assessment (EIA). It is worth quoting its objectives:

the best environmental policy consists in preventing the creation of pollution or nuisance at source, rather than subsequently trying to counteract their effects ... [involving] the need to take effects on the environment into account at the earliest possible stage in all the technical planning and decision making processes ... general principles for the assessment of environmental effects should be introduced with a view to supplementing and coordinating development

consent procedures governing public and private projects likely to have a major effect on the environment.... [T]he effects of a project on the environment must be assessed in order to take account of concerns to protect human health, to contribute by means of a better environment to the quality of life, to ensure maintenance of the diversity of species and to maintain the reproductive capacity of the ecosystem as a basic resource for life.
(Quoted in Hawke 1987:1)

If objectives such as these could be converted into workable procedures and specific applications in transport, then we could well see for the first time a comprehensive evaluation of the effects of, for example, new roads. This could include land-take, ecological damage, road traffic accidents, and effects of pollution on human health, together with many more factors currently excluded from consideration.

For the purposes of the Directive, 'project' is taken to mean two things: (a) the execution of construction works or of other installations or schemes, and (b) other interventions in the natural surroundings and landscape including those involving the extraction of mineral resources. Motorways are listed as an example under the first heading, while urban road construction is not.

Conclusion

Motorization, with its multi-faceted, ramifying environmental impact, is a major source of disturbance, nuisance, health hazards, land-take, and ecological disruption. The EEC in its environmental policies has identified a narrow sub-set of this range for detailed attention, failing to identify the interrelated issues of land-take, community disruption, psychological disturbance, and conservation threats in urban areas which arise from road construction programmes. At larger geographical scales, the EEC has failed to identify the links between motorization, resource consumption, energy depletion, and the creation of waste. Likewise, it has failed to grasp the significance of public transport, walking, and cycling as attractive environmental policies in addition to their transport advantages. In its infrastructure policies, and in its stance on the de-regulation of air transport, the EEC may also be contributing to environmental problems (though these fade into relative insignificance when compared with agricultural and nuclear policies of the Community). Air deregulation has the potential to make serious inroads into rail traffic, with serious implications for noise levels and land-take around airports, pollution, and the loss of rail services due to competition which, in its turn, can precipitate even greater levels of private transport trips between cities up to 200 km apart. In these respects, EEC environmental policies have failed both to grasp the enormity of the threat to the environment posed by motorization and the threat posed by the Commission's own policies in cognate areas.

12 Conclusion

It is not the purpose of a conclusion to rehearse what has gone before, but rather to give emphasis and clarity to the main points to emerge from the discussion. There can be little doubt that the Common Transport Policy has been an imperfect instrument of EEC policy. Its historical development, and its frequent reviews of what has gone before and what is to come next, have the unconvincing aura of a bureaucratic process in search of a role! This may not be a bad thing. An effective policy rooted in liberalization and 'freeing the market' may well produce adverse effects as critics of the Common Agricultural Policy would claim.

The CTP has no clear vision of transport as a process and an important determinant of economic and social justice, environmental quality and the quality of life in its widest sense. The CTP views transport as a crude product, the consumption of which can be increased with little thought for the consequences. Moreover, such increased consumption is seen as a valid objective in itself, indicating greater output and greater progress towards transport at least cost to the consumer.

Liberalization is primarily about the goals of efficiency, cost minimization and eliminating bureaucratic controls as at frontiers. This is too crude. Cost minimization cannot take place in a vacuum. There will be transfers of costs from country to country or private to public interests as the overall transport system adjusts to the periodic shocks which the evolving CTP generates. Benefits will be created and destroyed, disbenefits will be generated and redistributed. What is striking about the CTP is that it has such an impressively simplistic concept of transport with no awareness of disbenefits whether environmental, infrastructural, or social.

The attitude of the EEC to European public transport, the railways, the cities, cars, and pedestrians and cyclists has been described in this book. EEC actions in these policy areas were found wanting. This is a disappointment and represents a lost opportunity for the improvement of living and working conditions in Europe.

In many ways, the CTP is merely a pale reflection of what happens at the level of individual countries. With some notable exceptions in Germany, the Netherlands, and Denmark, the general approach to major transport problems within an individual country displays many of the same characteristics. More roadbuilding, sustained high levels of road traffic accidents, and atmospheric pollution, and the hypermobility of the wealthy compensated for by the immobility of the poor, all can be found in every European country. The accession of Greece, Spain, and Portugal does not fundamentally alter this picture, but does put these countries on a similar trajectory to that being experienced by the more 'advanced' nations. The CTP is a major force for 'legitimating' a particular view of transport and for giving this one position out of many considerable impetus with its legal instruments and financial incentives. In this respect, a supra-national body has a much more responsible role than its national equivalent. In the case of the EEC, the transport decisions and transport politics of twelve countries are directly influenced by the stance taken by this larger body. Not only are new members pressurized on lorry weights, railway finances, road construction, etc., but for existing members the possibility of an alternative view emerging from the political analysis·of transport decisions is reduced by the weight of the legal instruments of the supra-national body. This destroys diversity and gives alternative views little room to gain credibility.

A large institution does have the option of moving the transport debate along different lines and indeed subscribing to a fundamental analysis of objectives and the means of attaining objectives. The EEC has not done this and more generally reflects the prevailing balance of power and advantage within the transport world - a situation very succinctly described by Hamer (1987).

The CTP stands, therefore, as an ineffective policy pursued in an indifferent manner based on simplistic economic notions.

The latest proposals in 'freeing the market' show that the CTP is poised to enter a vigorous phase, and one that points to greater use of road vehicles for freight and international freight. The consequences of this have not been spelt out in detail, but will certainly include a renewed demand for more road space, an increased environmental disturbance, and damage to the fabric of towns and the ordinary life of those towns. This is what is at stake in a transport policy, and this is where the evidence points to a strong negative effect of the Common Transport Policy.

Appendix: TIEP Projects

The following is a list of projects submitted for consideration to the EEC Commission as part of the Transport Infrastructure Experimental Programme. *Source*: Transport Infrastructure Experimental Programme, Commission of the European Communities, Brussels. Communication from Commission to Council, COM(82) 828 Final, 10th December, 1982.

A. Main communication routes within the Community

Mode of Transport	Axis or Location	Type of Project	Estimated Cost (m. ECU)
Rail	Brussels–Namur Luxembourg French border towards Metz	Increases in capacity and speed in particular by:	
		– third line on a section	1,500
		– straightening of line, signalization work and intensification of overhead power line (Belgium–Luxembourg)	not yet costed

	Axis	Location	
Rail	Paris–Brussels–Aachen Cologne Axis	Construction of a new line enabling the running of TGV (French section)	600
		Brussels–Aachen; third line, straightening of line, signalization (Belgian section)	not yet costed
		Aachen–Cologne (FRG) Rapid removal of bottlenecks Improvement of service level	42
Rail	north/south axis Athens–Thessalonika–Idomeni (Yugoslav border)	Electrification and line improvement work	366
Rail	Athens–Korinthos Patras and Korinthos–Argos axes	Resignalling	25
Rail	north/south axis. Helsingør Copenhagen–Milan	Electrification work and increase in the number of lines in Denmark	108
		Rapid removal of bottlenecks on the Hamburg–Lübeck line (FRG)	42
		Construction of a third line on a section of the Hamburg–Lübeck line (FRG)	107

Mode of Transport	Axis or Location	Type of Project	Estimated Cost (m. ECU)
Rail	north/south axis. Helsingør Copenhagen–Milan (continued)	Construction of extra capacity on the Milan–Chiasso line (Gothard line) Italy	74.3
		Installation of automatic signalling on the Domodossola Gallarate section (Simplon line)	43.8
Rail	Amsterdam–Rotterdam–Cologne–Munich–Verona axis	Various projects intended to remove bottlenecks on the Amsterdam Rotterdam Breda Eindhoven Venlo line (fourth line – 4 line tunnel) heightening of draw-bridge (the Netherlands)	474
		Construction of extra capacity in tunnel on sections of the Brenner–Bolzano line	158.9
		Doubling of lines on certain sections of the Verona–Bologna line	
Rail	Rail junctions on main lines	Development of trans-shipment stations of Cologne, Eifeltor, Kornwestheim, Regensburg Ost	114

Road	north–south axis Amsterdam–Liège Luxembourg–Saarbrücken (E 25, E 420 and E 27 roads)	Construction of road between Maasbracht and Boxmeer	235
		Development of the Luxembourg–Ettelbrück road	130
		Development of the Luxembourg–German border road	43
	east–west axis Rotterdam–Eindhoven – FRG (E 25)	Weert By-pass	25
	Netherlands – FRG axis (E 30)	Construction of a section between Enschede and the border	29
Road	north–south axis	Development of the road in Denmark (construction of motorway sections, bridges, etc.)	223
	Copenhagen Hamburg Hannover Würzburg (E 45) Ulm	Widening of a section of the motorway between Hamburg and Hannover	75
	Memmingen towards Austria and Switzerland	Development of the lines Memmingen–Lindau and Memmingen towards Innsbruck	124

Mode of Transport	Axis or Location	Type of Project	Estimated Cost (m. ECU)
Road	Luxembourg – Trèves link	Construction of a missing link near the German border	38
Road	east-west axis (E 90 and E 950 roads)	Construction of motorway section in the Grand Duchy	136
Road		Development of the links between Igoumenitsa–Volos and Igoumenitsa Thessalonika	900–1,050
Road	north-south axis	Development of the Volos–Athens–Korinthos–Kalamata route	1,050
Road	Rosslare–Dublin Belfast axis (E 01 road)	Construction of various town by-passes on the road between Rosslare–Dundalk	100
Road	Simplon axis (E 62 road)	Modernization of the section Ornavasso Domodossola to the north of Milan	48.3
Road	Belgium/Netherlands/FRG via Aachen and Cologne	Widening of a section of motorway between Cologne and Aachen	36
Road	Rotterdam–Cologne axis (E 33 road)	Developments of certain points along the road (ring, bridge …)	23

Road	Ireland Continent axis via Holyhead, ports of Harwich, Dover, Folkestone, Southampton (E 22, E 05, E 15, E 28 and E 30 roads)	Construction or development of various sections, by-pass of built-up area (of which the E 15/E 30 roads avoid London)	837
Inland waterways	France–Belgium axis	Development of the Lys link	58
	east–west axis	Development of the canal du Centre (Belgium)	154
	Belgium Netherlands axis	Development of the Albert canal	264
		Development of the Belgium section of the Lanaye canal	22
		Development of the Dutch section of the Zuid-Willemsvaart canal and the Belgian section	60
		Development of the South Beveland canal (the Netherlands)	235
		Development of the Wessem–Nederweert (the Netherlands)	60
	Rhine axis	Deepening of the lower Rhine between Duisburg and the Dutch border	28

B. Transit routes between Community member states traversing third countries

Mode of Transport	Axis or Location	Type of Project	Estimated Cost (m. ECU)
Rail	Salzburg–Villach Rosenbach line	Widening of gauge and improvement of the track (in particular for combined transport)	84
	(Yugoslav border)	Removal of bottlenecks	
Road	north–west/south–east axis Nürnberg–Linz Graz–Zagreb	Construction of a section of the Innkreis motorway (Austria)	106

C. Communication routes within the Community, important for regional integration

Mode of Transport	Axis or Location	Type of Project	Estimated Cost (m. ECU)
Rail	Antwerp–Hasselt Maestricht–Monzen–Aachen link	Development of certain section to have a more direct line for passengers	23
Rail	Antwerp–Athus Longwy link	Electrification of sections	not yet costed
Rail	Maestricht–Liège Luxembourg	Electrification of the section Visé Kinkempois Gouvy, Luxembourg	not yet costed
Rail	Various lines in Greece	Resignalling	61
Road	Kalamata–Patras Igoumenitsa (E 55 road)	Bridge construction from Rio–Antirio	300
Road	Grosseto–Fano (E 78 road)	Construction of a section between Calmazzo and Bivio Bolzaga	19.3
Road	E 90 road Brindisi–Mazara del Vallo (Sicily) and E 45 road Salerno–Messina–Gela (Sicily)	Modernization of various sections	251.8

Mode of Transport	Axis or Location	Type of Project	Estimated Cost (m. ECU)
Road	Netherlands/FRG links; through the north (E 22) and links with E 23 and E 232	Development or construction of sections	164
Road	Stranraer–Newcastle (E 18) road. In Scotland E 16, E 15, A 36 and A 32 roads	Development of various sections	138
Road	Netherlands/Belgium link	Various developments in border regions Bridge over the Western Escaut	15 350–550
Inland waterway	Oude Maas	Construction of a draw-bridge	16

D. Main communication routes of importance for traffic between the Community and third countries

Mode of Transport	Axis or Location	Type of Project	Estimated Cost (m. ECU)
Rail	Thessaloniki Alexandroupolis Ormenio Axis (towards Turkey)	Development of the line	97
Road	Thessaloniki Turkey axis	Development of the road between Thessaloniki and Greek–Turk border	120–135
Rail	Denmark Sweden axis	Development of the line between Copenhagen and Rodby (electrification and increase in the number of tracks in Denmark)	(already accounted for in table A)
Road	Denmark Sweden axis	Development of the E 45 road between Copenhagen and Rodby	(already accounted in table A)

E. Access routes to ports and airports of importance for traffic between member states or traffic between the Community and third countries. Installations in these ports or airports

Mode of Transport	Axis or Location	Type of Project	Estimated Cost (m. ECU)
Rail	Colchester Harwich line	Electrification	47
	London Gatwick line	Improvement of the facilities at London Victoria station	42
	Manchester airport line	New electrified line	55–64
Airport and air control installations	Greece	Development of various airports Modernization of the air control system	103
Airport	Ireland	Development of the airports of Cork, Shannon, Charlestown (construction)	25
Port	Ireland	Development of Waterford port	3

Port	United Kingdom	Various developments, concerning in particular combined transport at the ports of Dover, Harwich, Portsmouth, Felixstowe and Great Yarmouth (cost not known)	75
Airport and air control installation	United Kingdom	Various developments at Gatwick, Manchester, Liverpool, Belfast, Edinburgh and London airports	383

References

Abbess, C., Jarrett, D., and Wright, C.C. (1981) 'Accidents at blackspots: estimating the effectiveness of remedial treatment, with special reference to the 'regression-to-mean' effect', *Traffic Engineering and Control*, 22(10), 535–42.

Adams, J. (1981) *Transport Planning: Vision and Practice*, London, Routledge & Kegan Paul.

Adams, J. (1985) *Risk and Freedom – the Catalogue of Road Safety Regulation*, Cardiff, Transport Publishing Projects.

Allsop, R. (1983) 'Fares and road casualties in London', *Report of the Transport and Development Department of the GLC*, University of London Transport Studies Group.

Allsop, R.E. and Turner, E.D. (1984) 'Road casualties and public transport fares in London', paper presented at the International Workshop on the Methodology of Modelling Road Accident and Injury Patterns, University of Sussex, 9–10 July.

Appleyard, D. (1981) *Liveable Streets*, Berkeley, Calif., University of Californian Press.

Armitage, A. (1981) *Lorries, People and the Environment*, Cmnd 8439, London, HMSO.

Baughan, C.J., Hedges, B., and Field, J. (1983) 'A national survey of lorry nuisance', *Transport and Road Research Laboratory Supplementary Report No. 774*.

Bayliss, B. (1975) 'Raw material resources and transport' in *Sixth International Symposium on Theory and Practice in Transport Economics*, Madrid, European Conference of Ministers of Transport pp. 107–44,

Bayliss, B.T. (1979) 'Transport in the European Communities', *Journal of Transport Economics and Policy* 13, 1, 28–43.

Bayliss, B.T. and Edwards, S.L. (1970) *Industrial Demand for Transport*, Ministry of Transport, London, HMSO.

Beardwood, J. and Elliot, J. (1985) 'Roads generate traffic', London, Greater London Council (mimeo.)

214

Blumer, W. and Reich, T. (1980) 'Leaded gasoline - a cause of cancer', *Environment International,* 3, 465-71.

Boyle, A.J. and Wright, C.C. (1984) 'Accident "migration" after remedial treatment at accident blackspots' (1984) *Traffic Engineering and Control,* 25(5), 260-166.

Bundesminister für Verkehr (1987) 'Mitteilungen Über Forschungen, zur Verbesserung der Verkehrsverhältnisse der Gemeinden', *Forschung Stadtverkehr Heft 40.*

Button, K.J. (1984) *Road Haulage Licensing and EEC Transport Policy,* Aldershot, Gower.

Channel Tunnel Joint Consultative Committee (1986) *Kent Impact Study: A Preliminary Assessment,* London, CTJCC.

Chini, L.W. (1980) 'Competitive position and future of inland waterway transport', *Round Table* (ECMT), 49, 5-76.

Conservation Society (nd) *Common Transport Policy,* London, Conservation Society.

Couper, A.D. (1977) 'Shipping policies of the EEC', *Maritime Policy and Management,* 4, 129-39.

Cumbria County Council (1987) *Settle-Carlisle Railway: The Case against Closure,* Cumbria C.C.

Davies, D.G. (1984) 'A survey of county council cycle planning in Britain', *Traffic Engineering and Control,* 25(5), 182-5.

Department of Transport (1981) *Cycling: A Consultation Paper,* London, HMSO.

Department of Transport (1983) *National Travel Survey: 1978/79 Report,* London, HMSO.

Department of Transport (1984) *Road Accidents: Great Britain, 1983,* London, HMSO.

Despicht, N. (1969) *The Transport Policy of the European Communities,* European Series No. 12 (PEP), London, Chatham House.

Dix, M.C. and Layzell, A.D. (1983) *Road Users and the Police,* London, Croom Helm.

Dodgson, J.S. (1984) 'Railway costs and closures', *Journal of Transport Economics and Policy,* 18, 3, 219-35.

EC Commission (1961) *Memorandum on the General Lines of a Common Transport Policy* (Schaus Memorandum).

EC Commission (1976) 'Environment Programme 1977-81', *Bulletin of the European Community,* Suppl. 6.

EC Commission (1977) *Report of an Inquiry into the Current Situation in the Major Community Seaports,* Port Working Group.

EC Commission (1979a) *A Transport Network for Europe: Outline of a Policy*, Bulletin 8/79.

EC Commission (1979b) '*Report on Possible Measures to Rehabilitate Inland Waterway Transport*', Document 146/79, 7 May.

EC Commission (1981) *Proposal for a Council Regulation Laying Down Detailed Rules for Application of Articles 85 and 86 of the Treaty to Maritime Transport*, 10150/81 (+ADD 1), COM (81) 423 Final.

EC Commission (1982) *The Community and the Countries and Regions of the Mediterranean*, European File 19/82, December.

EC Commission (1983a) *Carossino Report on Ports*, OJ C96 11.4.83.

EC Commission (1983b) *Commission Communication to the Council 'Progress towards a Common Transport Policy (Inland Transport)'*, COM (83) Final.

EC Commission (1985a) *The Community and Transport Policy*, European File 10/85.

EC Commission (1985b) *European Transport Annual Report, 1985*.

EC Commission (1985c) 'Progress towards a Common Transport Policy (Maritime Transport)', *Bulletin of the European Communities*, Suppl. 5/85, 5635/85, COM (85) 90.

EC Commission (1987) *The European Community and Environmental Protection*, European File 5/87.

ECMT (1985) *31st Annual Report - 1984*, Activity of the Conference, Resolution of the Council of Ministers of Transport and Reports Approved in 1984, Paris, OECD.

ECMT (1986a) *32nd Annual Report - 1985*, Activity of the Conference, Resolutions of the Council of Ministers of Transport and Reports Approved in 1985, Paris, OECD.

ECMT (1986b) *Principal Activities of ECMT in the Field of Road Safety*, Paris, OECD.

Erdmenger, J. (1983) *The European Community Transport Policy*, Aldershot, Gower.

European Parliament (1986) *Committee on Transport, Draft Report on the bicycle as a means of transport. Rapporteur, F. Wijsenbeek, Part B: Explanatory Statement*, WG (VS1) 471 9E PE (107.416/B).

Eurostat (1985) *Statistical Yearbook, Transport, Communications, Tourism 1970-1983*, Brussels.

Federal Republic of Germany (FRG) (1984) *Road Safety Programme, 1984, of the Federal Government, Bonn*, Federal Ministry of Transport, Bonn, West Germany.

Frohnmeyer, A. (1983) *Speech to Transport 2000 26/9/83*, London, The Future of Community Railways.

Frybourg, M. (1984) 'France' in *The Costs of Combined Transport*, ECMT Round Table 64, Paris, ECMT, pp. 5–30.

FTA (1986) *EEC – the Free Movement of Goods: The Liberalisation of the European Freight Transport Market.*

Gakenheimer, R. (1978) *The Automobile and the Environment: An International Perspective*, Transportation Studies series, No. 1, Cambridge, Mass., MIT Press.

Greater London Council (GLC) (1983) *Thirty Years On – a Review of Air Pollution in London*, London, GLC.

GLC (1986a) *The Adequacy of British Rail's London Services*, London, GLC.

GLC (1986b) *London Transport Policy and Programme*, London, GLC.

Grimshaw, J.R. (1982) 'Common opportunities – notes on the experience of a cycling campaigning group', paper presented at the Facets of a National Cycling Policy Conference, Bristol University.

Gwilliam, K. (1980) 'Realism and the Common Transport Policy of the EEC' in J.B. Palak and J.B. Van der Kamp, (eds) *Changes in the Field of Transport Studies*, The Hague, Martinus Nijhoff.

Haigh, N. and Hand, C. (1983) 'Assessment of society's transport needs: goods transport' in *Transport Is for People*, Ninth ECMT International Symposium on Theory and Practice in Transport Economics, Madrid, 1982.

Hall, P. and Hass-Klau, C. (1985) *Can Rail Save the City?*, Aldershot, Gower.

Hamer, M. (1987) *Wheels within Wheels: A Study of the Road Lobby*, London, Routledge & Kegan Paul.

Haupt, R. and Wilken, D. (1985) 'Luftverkehrsnachfrage und Neubauplan der Bundesbahn Substitutionseffekt', *Internationales Verkehrswesen*, 37, 405–9.

Hawke, N. (1987) *Environmental Impact Assessment North American and European Developments*, Leicester Polytechnic Law School Monograph.

Haworth, C.I. (1985) 'Perceived safety and strategies for reducing accidents' in *'Evaluation 85' International Meeting on the Evaluation of Local Traffic Safety Measures, Paris, 20–23 May 1985*, eds M.B. Biecheter, C. Lacombe, and N. Muhlrad, BP 34 94114 Arcueil, Cedex, France, 445–452.

Hillman, M. and Whalley, A. (1977) *Fair Play for All: A Study of Access to Sport and Informal Recreation*, Political and Economic Planning Broadsheet No. 571, London.

Hillman, M. and Whalley, A. (1979) *Walking Is Transport*, London, Policy Studies Institute.

Hillman, M., Henderson, I., and Whalley, A. (1976) *Transport Realities and Planning Policy*, Political and Economic Planning Broadsheet No. 567, London.

Holman, C. (1987) *Air Pollution from Diesel Vehicles*, London, Friends of the Earth.

Holzapfel, H. (nd) *High-speed Systems of Public Transport - a Positive Trend?*, Berlin, Technische Universität.

Hoppner, M. (1979) 'Fahrradverkehr als Beitrag zur prinzipiellen Verkehrsberuhigung' in Verkehr in der Sackgasse: kritik und alternativen', *Technologie und Politik*. Das Magazin zur Wachstumstail, Rowohlt Taschenbuch Verlag GMBH, 175-204.

House of Commons (1985) *Road Safety. First Report of the Transport Committee. Session 1984-85*, 103.I, 103.II, 103.III, London, HMSO.

House of Lords (1983a) *Select Committee on the European Community. Competition Policy: Shipping. Session 1983-84*, Third Report, 23.

House of Lords (1983b) *EEC Social Regulations. Session 1983/84*, HL 221.

House of Lords (1985a) *Select Committee on the European Community. Session 1984-85*, Seventh Report European Air Transport Policy, HL 115.

House of Lords (1985b) *Select Committee on the European Community. Lead in Petrol and of Vehicle Emissions. Session 1984-85*, Fifth Report, HL 96.

Hudson, M. (1982) *Bicycle Planning, Policy and Practice*, London, Architectural Press.

Janssen, S.T.M.C. (1985) 'Effects of road safety measures on urban areas' in ONSER, *'Evaluation 85' International Meeting on the Evaluation of Local Traffic Safety Measures, Paris, 20-23 May 1985*, eds M.B. Biecheter, C. Lacombe, and N. Muhlrad, Paris, ONSER, 168-83.

Jeschke, C. and Kunert, U. (1984) *Die Zukunft des Automobils. Teil D, Benutzerverhalten*, Berlin, Technische Universität.

Joint Unit for Research in the Urban Environment (1976) *Environmental Effect of Urban Road Traffic*, Report of Study, Birmingham, University of Aston.

Klatte, E. (1986) 'The past and the future of European environmental policy', *European Environment Review*, 1(1), 32-4.

Kracke, R., Aberle, G., Engelmann, H., Schleth, J.P., and Aubry, C. (1986) 'Aufgaben und Schwerpunkte der Forschung auf dem Gebiet des spurgeführren Schnellverkehrs vor dem Hintergrund technischer und okonomischer Perspecktiven Westeuropas', *Internationales Verkehrswesen*, 38, 7-10.

Kutter, E. (1987) 'Dilemmas in car mobility, the disenchantment of speed' in P. Nijkamp and S. Reichman (eds) *Transportation Planning in a*

Changing World, Aldershot, Gower, 48–60.

Lancaster City Council (1987) *Lancaster Local Plan*, Lancaster, Lancaster C.C.

Land Use Consultants (1985) *Channel Fixed Link: Environmental Appraisal of Alternative Proposals*, London, LUC.

London Borough of Brent (1986) 'Tesco, Neasden: a study of the retail impact' Brent Planning Department.

London Strategic Policy Unit Transport Group (1987) *The Channel Tunnel and London: An Examination of British Rail's Plans for London Terminal Facilities*, London, LSPUTG.

McGowan, F. and Trengove, C. (1986) *European Aviation – a Common Market?*, London, Institute for Fiscal Studies.

McShane, M. (1984) *The Future of the Automobile, Part D*, Cambridge, Mass., MIT Press.

Nash, C. (1985) 'European railway comparisons – what can we learn?' in K.J. Button and D.E. Pitfield (eds) *International Railway Economics*, Aldershot, Gower, pp. 237–70.

National Audit Office (NAO) (1985) *Report by the Comptroller and Auditor General. Department of Transport: Expenditure on Motorways and Trunk Roads*, London, NAO.

Nilsson, G. (1981) 'The effects of speed limits on traffic accidents in Sweden' in *OECD Symposium, 'The Effects of Speed Limits on Traffic Accidents and Transport Energy Use', Dublin, 6–8 October*, Paris, OECD.

North Rhine Westphalia (1979) 'Verkehrsberuhigung in Wohngebieten', Der Minister für Wirtschaft, Mittelstand und Verkehr.

Oliver, K. (1988) *Different Accessibility and Transport in Britain*, Contemporary Issues in Geography and Education (forthcoming).

Organisme National de Sécurité Routière (ONSER) (1985) *'Evaluation 85' International Meeting on the Evaluation of Local Traffic Safety Measures, Paris, 20–23 May 1985*, eds M.B. Biecheter, C. Lacombe, and N. Muhlrad, Paris, ONSER.

Organization for Economic Co-operation and Development (OECD) (1981) 'The effects of speed limits on traffic accidents and transport energy use' *Proceedings International Symposium Dublin, An Foras Forbatha, Dublin*, Paris, OECD.

OECD (1983a) *Traffic Safety of Children*, Paris, OECD.

OECD (1983b) *Road Research: Impact of Heavy Freight Vehicles*, Report prepared by an OECD road research group, December 1982, Paris, OECD.

Pickup, L. and Town, S.W. (1983) *Commuting Patterns in Europe: An*

Overview of the Literature, Transport and Road Research Laboratory Report, Supplementary Report 796.

Plowden, S. (1985) *Transport Reform: Changing the Rules*, London, Policy Studies Institute.

Pryke, R.W.S. and Dodgson, J. (1975) *The Rail Problem*, Oxford, Martin Robertson.

Retzko, H.G. and Sturm, P. (1986) 'Was nun? Was tun? Einige Bemerkungen zur Lage nach dem Abgas Grossversuch auf Autobahnen', *Internationales Verkehrswesen*, 38, 11–16.

Roberts, J. (1981) *Pedestrian Precincts in Britain*, London, TEST.

Roberts, J. (1987) 'Transport planning in continental cities – lessons for Britain', paper presented at TCPA Conference on 13th May 1987, the Car, Planning, and the Environment.

Roberts, Jonathan (1987) 'Bottlenecks or opportunities: the impact of the Channel Tunnel on railway opportunities', *Transport Report*, 10(4), 4–5.

RoSPA (Royal Society for the Prevention of Accidents) (1987) 'This old bridge is falling down ... and this ...', *Care on the Road*, May p. 2.

Rumar, K. (1985) 'Safety Problems and countermeasure effects in the Nordic countries' in ONSER, *'Evaluation 85' International Meeting on the Evaluation of Local Traffic Safety Measures, Paris, 20–23 May 1985*, eds M.B. Biecheter, C. Lacombe, and N. Muhlrad, Paris, 151–67.

Rylander, R., Ahrlin, U., and Vesterlund, J. (1978) Medical effects of environmental noise exposure, Gothenburg, Sweden, 1977, *Journal of Sound and Vibration*, 59(1), 59–142.

Sciarrone, G. and Carrara, M. (1984), 'Italy' in *The Costs of Combined Transport*, Report of Sixty-fourth Round Table on Transport Economics, Paris, ECMT, pp. 51–76.

Secrett, C. and Hodges, V.C. (1986) *Motorway Madness. Roads and their Impact on the Natural Environment*, London, Friends of the Earth.

Seidelmann, C. (1984) 'Germany' in *The Costs of Combined Transport*, ECMT Round Table No. 64, Paris, ECMT, pp. 31–50.

Serpell, D. (1983) *Railway Finances*, London, HMSO.

Simmons, M. (1987) 'The impact of the Channel Tunnel', paper presented at Transport Geography Study Group Annual Conference of the Institute of British Geographers, Portsmouth, 6th January 1987.

Simon, D. (1987) 'Spanning muddy waters: the Humber Bridge and regional development', *Regional Studies*, 21(1), 25–30.

Steer, Davies, and Gleave, (1987) *Turning Trucks into Trains: The Environ-*

mental Benefits of the Channel Tunnel, London, Transport 2000.

TEST (1984) *BR: A European Railway*, London, TEST.

TEST (1985) *The Accessible City*, London, TEST.

TEST (1986) *Re-training the Settle-Carlisle Line*, London, TEST.

TEST (1988) *Quality Streets: How Traditional Urban Centres Benefit from Traffic-Calming*, London, TEST.

Tortrat, C. (1985) 'Une Etude de cas en France: la sécurité routière au Mans' in ONSER, *'Evaluation 85' International Meeting on the Evaluation of Local Traffic Safety Measures, Paris, 20–23 May 1985*, eds M.B. Biecheter, C. Lacombe, and N. Muhlrad, Paris, ONSER.

Transport 2000 (1985) *Cities before Roads*, Evidence to Review of Urban Road Appraisal Methods, London, Transport 2000.

Transport 2000 (1986) *Getting There Alive: New Ideas in Road Safety*, Conference held at Kensington Town Hall, London, 20 March, 1986, London, Transport 2000.

Twitchett, C.G. (1981) 'Harmonisation and road freight transport' in C.G. Twitchett (ed.) *Harmonisation in the EEC*, London, Macmillan. Ch. 5 pp. 63–77.

Union Internationale des Chemins de Fer (UIC) (1985) *International Railway Statistics*, Paris, UIC.

Van der Bos, M. (1983) 'Management of firms to satisfy transport needs, inland waterways goods transport' in *'Transport is for People'*, *Ninth International Symposium on Theory and Practice in Transport Economics, Madrid, 2–4 November 1982*, European Conference of Ministers of Transport, OECD, Paris.

Vanke, J. (1986) *Better Roads for a Better Economy? A Literature Review*, London, Friends of the Earth.

Verband der Automobilindustrie, (VDA) (1985) *Fakten gegen Tempolimit auf Autobahnen*, Frankfurt am Main, VDA.

Vickerman, R.W. (1987) 'The Channel Tunnel: consequences for regional growth and development', *Regional Studies*, 21(3), 187–98.

Watkins, S.M. (1984) *Cycling Accidents*, Final Report on a survey of cycling and accidents, Godalming, Cyclists' Touring Club.

Whitelegg, J., (1979) 'The Common Transport Policy: a case of lost direction', *Transportation Science*, 13(4), 343–57.

Whitelegg, J. (1983), 'Road safety: defeat, complicity and the bankruptcy of science', *Accident Analysis and Prevention*, 15(2), 153–60.

Whitelegg, J. (1984a) 'Closure of the Settle–Carlisle railway line: the case for a social cost–benefit analysis', *Land Use Policy*, 1(4), 283–98.

Whitelegg, J. (1984b) 'The company car in the UK as an instrument of transport policy', *Transport Policy and Decision Making*, 2, 219–30.

Whitelegg, J. (1985) 'Road building and the urban economy' in *Cities and Roads, Proceedings of a Conference Held at the City University, London, 26 November 1985*, London, Transport 2000.

Whitelegg, J. (1987a) 'The destruction of a transport system', paper presented at Annual Conference, Institute of British Geographers, Portsmouth, 5th January, 1987.

Whitelegg, J. (1987b) 'A geography of road traffic accidents', *Transactions Institute of British Geographers*, 12(1), 161–76.

Wood, D. (1983) *Heavy Lorries in London*, London, GLC.

Yago, G. (1984) *The Decline of Transit: Urban Transportation in German and US cities, 1900-1970*, Cambridge, Cambridge University Press.

Yeowart, N.S., Wilcox, D.J., and Rossall, A.W. (1977) 'Community reactions to noise from freely flowing traffic, motorway traffic and congested traffic flow', *Journal of Sound Vibration*, 53(1), 127–45.

Index

For Product Safety Concerns and Information please contact our EU
representative GPSR@taylorandfrancis.com
Taylor & Francis Verlag GmbH, Kaufingerstraße 24, 80331 München, Germany